www.EffortlessMath.com

... So Much More Online!

- ✓ FREE Math lessons

- ✓ More Math learning books!

- ✓ Mathematics Worksheets

- ✓ Online Math Tutors

Need a PDF version of this book?

Visit www.EffortlessMath.com

ISEE Lower Level Mathematics Prep 2019

A Comprehensive Review and Ultimate Guide to the ISEE Lower Level Math Test

By

Reza Nazari & Ava Ross

All inquiries should be addressed to:

info@EffortlessMath.com

www.EffortlessMath.com

ISBN-13: 978-1-970036-10-7

ISBN-10: 1-970036-10-9

Published by: Effortless Math Education

www.EffortlessMath.com

Description

ISEE Lower Level Mathematics Prep 2019 provides students with the confidence and math skills they need to succeed on the ISEE Lower Level Math, building a solid foundation of basic Math topics with abundant exercises for each topic. It is designed to address the needs of ISEE Lower Level test takers who must have a working knowledge of basic Math.

This comprehensive book with over 2,500 sample questions and 2 complete ISEE Lower Level tests is all you need to fully prepare for the ISEE Lower Level Math. It will help you learn everything you need to ace the math section of the ISEE Lower Level.

Effortless Math unique study program provides you with an in-depth focus on the math portion of the exam, helping you master the math skills that students find the most troublesome.

This book contains most common sample questions that are most likely to appear in the mathematics section of the ISEE Lower Level.

Inside the pages of this comprehensive ISEE Lower Level Math book, students can learn basic math operations in a structured manner with a complete study program to help them understand essential math skills. It also has many exciting features, including:

- Dynamic design and easy-to-follow activities
- A fun, interactive and concrete learning process
- Targeted, skill-building practices
- Fun exercises that build confidence
- Math topics are grouped by category, so you can focus on the topics you struggle on
- All solutions for the exercises are included, so you will always find the answers
- 2 Complete ISEE Lower Level Math Practice Tests that reflect the format and question types on ISEE Lower Level

ISEE Lower Level Mathematics Prep 2019 is an incredibly useful tool for those who want to review all topics being covered on the ISEE Lower Level test. It efficiently and effectively reinforces learning outcomes through engaging questions and repeated practice, helping you to quickly master basic Math skills.

About the Author

Reza Nazari is the author of more than 100 Math learning books including:
– **Math and Critical Thinking Challenges:** For the Middle and High School Student
– **ACT Math in 30 Days.**
– **ASVAB Math Workbook 2018 – 2019**
– **Effortless Math Education Workbooks**
– **and many more Mathematics books …**

Reza is also an experienced Math instructor and a test–prep expert who has been tutoring students since 2008. Reza is the founder of Effortless Math Education, a tutoring company that has helped many students raise their standardized test scores—and attend the colleges of their dreams. Reza provides an individualized custom learning plan and the personalized attention that makes a difference in how students view math.

You can contact Reza via email at:
reza@EffortlessMath.com

Find Reza's professional profile at:
goo.gl/zoC9rJ

Contents

Chapter 1: Place Vales and Number Sense

Topics that you'll learn in this chapter:

✓ Place Values

✓ Compare Numbers

✓ Numbers in Numbers

✓ Rounding

✓ Odd or Even

Place Values

<table>
<tr><td rowspan="2">Helpful

Hints</td><td colspan="2">The value of the place, or position, of a digit in a number. For the number 3,684.26</td><td>Example:
In 456, the 5 is in "tens" position.</td></tr>
</table>

Decimal Place Value Chart

Millions	Hundred thousands	Ten thousands	Thousands	Hundreds	Tens	Ones	Decimal point	Tenths	Hundredths	Thousandths	Ten-thousandths	Hundred thousandths	Millionths
			3	6	8	4	.	2	6				

🖊 *Write numbers in expanded form.*

1) Thirty–five   30 + 5

2) Sixty–seven   ___ + ___

3) Forty–two   ___ + ___

4) Eighty–nine   ___ + ___

5) Ninety–one   ___ + ___

🖊 *Circle the correct choice.*

6)   The 2 in 72 is in the   ones place   tens place   hundreds place

7)   The 6 in 65 is in the   ones place   tens place   hundreds place

8)   The 2 in 342 is in the   ones place   tens place   hundreds place

9)   The 5 in 450 is in the   ones place   tens place   hundreds place

10)   The 3 in 321 is in the   ones place   tens place   hundreds place

Comparing and Ordering Numbers

Helpful *Hints*	**Comparing:** Equal to = Less than < Greater than > Greater than or equal ≥ Less than or equal ≤	To compare numbers, you can use number line! As you move from left to right on the number line, you find a bigger number!	**Example:** 56 > 35 Order integers from least to greatest. (− 11, − 13, 7, − 2, 12) − 13 <− 11< − 2 < 7 <12

✍ Use less than, equal to or greater than.

1) 23 _____ 34

2) 89 _____ 98

3) 45 _____ 25

4) 34 _____ 32

5) 91 _____ 91

6) 57 _____ 55

7) 85 _____ 78

8) 56 _____ 43

9) 34 _____ 34

10) 92 _____ 98

11) 38 _____ 46

12) 67 _____ 58

13) 88 _____ 69

14) 23 _____ 34

✍ Order each set numbers from least to greatest.

15) − 15, − 19, 20, − 4, 1   ___, ___, ___, ___, ___, ___

16) 6, − 5, 4, − 3, 2   ___, ___, ___, ___, ___, ___

17) 15, − 42, 19, 0, − 22   ___, ___, ___, ___, ___, ___

18) 26, − 91, 0, − 13, 67, − 55   ___, ___, ___, ___, ___, ___

19) − 17, − 71, 90, − 25, − 54, − 39   ___, ___, ___, ___, ___, ___

20) 98, 5, 46, 19, 77, 24   ___, ___, ___, ___, ___, ___

Write Numbers in Words

Helpful	- First, learn to write numbers from 1 to 9. 1 = one, 2 = two, 3 = three, 4 = four, 5 = five, 6 = six, 7 = seven, 8 = eight, 9 = nine
Hints	- Learn how to write numbers from 10 to 100. For example: 16 = sixteen, 50 = fifty, 89 = eighty-nine - Learn how to combine words to write more numbers. For example: 120 is written as: one hundred twenty

✍ *Write each number in words.*

1) 434 _____

2) 809 _____

3) 730 _____

4) 272 _____

5) 266 _____

6) 902 _____

7) 1,418 _____

8) 1,365 _____

9) 3,374 _____

10) 2,486 _____

11) 7,671 _____

12) 6,290 _____

13) 3,147 _____

14) 5,012 _____

Rounding Numbers

Helpful *Hints*	Rounding is putting a number up or down to the nearest whole number or the nearest hundred, etc.	**Example:** 64 rounded to the nearest ten is 60, because 64 is closer to 60 than to 70.

✍ *Round each number to the underlined place value.*

1) 1972

2) 2,995

3) 3364

4) 1281

5) 2355

6) 1334

7) 1,203

8) 1457

9) 7484

10) 1914

11) 4239

12) 9,123

13) 3,452

14) 2569

15) 1,230

16) 7698

17) 9293

18) 5237

19) 2493

20) 2,923

21) 9,845

22) 4555

23) 6939

24) 9869

Odd or Even

Helpful *Hints*	**Even:** Any integer that can be divided exactly by 2 is an even number. **Odd:** Any integer that cannot be divided exactly by 2 is an odd number.	**Example:** −3, 1, 7 are all odd numbers − 12, 0, 8 are all even numbers

✎*Identify whether each number is even or odd.*

1) 12 _____   5) 99 _____   9) 94 _____

2) 7 _____   6) 55 _____   10)   14 _____

3) 33 _____   7) 34 _____   11)   22 _____

4) 18 _____   8) 87 _____   12)   79 _____

✎*Circle the even number in each group.*

13) 22, 11, 57, 13, 19, 47   15) 19, 35, 24, 57, 65, 49

14) 15, 17, 27, 23, 33, 26   16) 67, 58, 89, 63, 27, 63

✎*Circle the odd number in each group.*

17) 12, 14, 22, 64, 53, 98   19) 46, 82, 63, 98, 64, 56

18) 16, 26, 28, 44, 62, 73   20) 27, 92, 58, 36, 38, 72

Answers of Worksheets – Chapter 1

Place Values

1) 30 + 5

2) 60 + 7

3) 40 + 2

4) 80 + 9

5) 90 + 1

6) ones place

7) tens place

8) ones place

9) tens place

10) hundreds place

Comparing and Ordering Numbers

1) 23 less than 34

2) 89 less than 98

3) 45 greater than 25

4) 34 greater than 32

5) 91 equal to 91

6) 57 greater than 55

7) 85 greater than 78

8) 56 greater than 43

9) 34 equal to 34

10) 92 less than 98

11) 38 less than 46

12) 67 greater than 58

13) 88 greater than 69

14) 23 less than 34

15) −19, −15, −4, 1, 20

16) −5, −3, 2, 4, 6

17) −42, −22, 0, 15, 19

18) −91, −55, −13, 0, 26, 67

19) −71, −54, −39, −25, −17, 90

20) 5, 19, 24, 46, 77, 98

Word Names for Numbers

1) four hundred thirty-four

2) eight hundred nine

3) seven hundred thirty

4) two hundred seventy-two

5) two hundred sixty-six

6) nine hundred two

7) one thousand, four hundred eighteen

8) one thousand, three hundred sixty-five

9) three thousand, three hundred seventy-four

10) two thousand, four hundred eighty-six

11) seven thousand, six hundred seventy-one

12) six thousand, two hundred ninety

13) three thousand, one hundred forty-seven

14) five thousand, twelve

Roman Numerals

1) II	6) VII	11) IX
2) VI	7) III	12) XI
3) IV	8) I	13) VI
4) IX	9) V	14) XII
5) X	10) VIII	

Rounding Numbers

1) 2000	9) 7500	17) 9290
2) 3000	10) 1910	18) 5240
3) 3360	11) 4240	19) 2490
4) 1280	12) 9000	20) 2900
5) 2360	13) 3450	21) 10,000
6) 1330	14) 2600	22) 4560
7) 1200	15) 1200	23) 6900
8) 1460	16) 7700	24) 9870

Odd or Even

1) even	8) odd	15) 24
2) odd	9) even	16) 58
3) odd	10) even	17) 53
4) even	11) even	18) 73
5) odd	12) odd	19) 63
6) odd	13) 22	20) 27
7) even	14) 26	

Chapter 2: Adding and Subtracting

Topics that you'll learn in this chapter:

- ✓ Adding Two–Digit Numbers
- ✓ Subtracting Two–Digit Numbers
- ✓ Adding Three–Digit Numbers
- ✓ Adding Hundreds
- ✓ Adding 4–Digit Numbers
- ✓ Subtracting 4–Digit Numbers

Adding Two–Digit Numbers

Helpful	1– Line up the numbers. 2– Start with the ones place. 3– Regroup if necessary. 4– Add the tens place. 5– Continue with other digits.	**Example:** $1,349$ $+2,411$ $3,760$
Hints		

✎ *Find each sum.*

1)
$$\begin{array}{r} 50 \\ + 18 \\ \hline \end{array}$$

2)
$$\begin{array}{r} 32 \\ + 14 \\ \hline \end{array}$$

3)
$$\begin{array}{r} 45 \\ + 16 \\ \hline \end{array}$$

4)
$$\begin{array}{r} 12 \\ + 12 \\ \hline \end{array}$$

5)
$$\begin{array}{r} 43 \\ + 30 \\ \hline \end{array}$$

6)
$$\begin{array}{r} 34 \\ + 15 \\ \hline \end{array}$$

7)
$$\begin{array}{r} 89 \\ + 7 \\ \hline \end{array}$$

8)
$$\begin{array}{r} 63 \\ + 12 \\ \hline \end{array}$$

9)
$$\begin{array}{r} 90 \\ + 10 \\ \hline \end{array}$$

Subtracting Two–Digit Numbers

Helpful	1– Line up the numbers. 2– Start with the units place. (ones place) 3– Regroup if necessary. 4– Subtract the tens place. 5– Continue with other digits.
Hints	

Example:

$$5,397$$
$$- 2,416$$
$$\overline{2,981}$$

✍ *Find each difference.*

1)
$$\begin{array}{r} 32 \\ -15 \\ \hline \end{array}$$

2)
$$\begin{array}{r} 40 \\ -12 \\ \hline \end{array}$$

3)
$$\begin{array}{r} 67 \\ -17 \\ \hline \end{array}$$

4)
$$\begin{array}{r} 18 \\ -10 \\ \hline \end{array}$$

5)
$$\begin{array}{r} 59 \\ -16 \\ \hline \end{array}$$

6)
$$\begin{array}{r} 89 \\ -20 \\ \hline \end{array}$$

7)
$$\begin{array}{r} 78 \\ -21 \\ \hline \end{array}$$

8)
$$\begin{array}{r} 66 \\ -15 \\ \hline \end{array}$$

9)
$$\begin{array}{r} 87 \\ -45 \\ \hline \end{array}$$

Adding Three–Digit Numbers

Helpful	1– Line up the numbers.	**Example:**
	2– Start with the unit place. (ones place)	120
	3– Regroup if necessary.	$+\ 114$
Hints	4– Add the tens place.	
	5– Continue with other digits.	234

✍️*Find each sum.*

1)
```
  234
+  56
-----
```

2)
```
  523
+ 134
-----
```

3)
```
  345
+ 167
-----
```

4)
```
  460
+ 120
-----
```

5)
```
  432
+ 430
-----
```

6)
```
  235
+ 150
-----
```

7)
```
  789
+  57
-----
```

8)
```
  863
+ 340
-----
```

9)
```
  956
+  89
-----
```

Adding Hundreds

Helpful *Hints*	1– Line up the numbers. 2– Start with the ones place. 3– Regroup if necessary. 4– Add the tens place and regroup if necessary. 5– Add the hundred digits.	**Example:** $\begin{array}{r} 500 \\ +\ 200 \\ \hline 700 \end{array}$

✐*Add.*

1) 200 + 200 = ——

2) 300 + 200 = ——

3) 500 + 200 = ——

4) 900 + 100 = ——

5) 100 + 700 = ——

6) 500 + 100 = ——

7) 200 + 800 = ——

8) 800 + 100 = —

9) 700 + 100 = —

10)   100 + 300 = ——

11)   500 + 500 = ——

12)   400 + 400 = ——

13)   300 + 400 = ——

14)   500 + 300 = ——

15)   If there are 600 balls in a box and Jackson puts 500 more balls inside, how many balls are in the box?

_____ balls

Adding 4–Digit Numbers

Helpful	1– Line up the numbers.	**Example:**
	2– Start with the unit place. (ones place)	
	3– Regroup if necessary.	$1,349$
Hints	4– Add the tens place.	$+2,411$
	5– Continue with other digits.	$3,760$

✍ *Add.*

1)
$$\begin{array}{r} 1,158 \\ + 6,687 \\ \hline \end{array}$$

3)
$$\begin{array}{r} 5,756 \\ + 2,712 \\ \hline \end{array}$$

5)
$$\begin{array}{r} 4,257 \\ +5,194 \\ \hline \end{array}$$

2)
$$\begin{array}{r} 5,188 \\ + 1,298 \\ \hline \end{array}$$

4)
$$\begin{array}{r} 3,239 \\ +2,562 \\ \hline \end{array}$$

6)
$$\begin{array}{r} 6,215 \\ +2,189 \\ \hline \end{array}$$

✍ *Find the missing numbers.*

7) $1,145 + \underline{\hphantom{xxx}} = 1,276$

10) $455 + \underline{\hphantom{xxx}} = 1,755$

8) $500 + 1,000 = \underline{\hphantom{xxx}}$

11) $\underline{\hphantom{xxx}} + 720 = 1,250$

9) $3,200 + \underline{\hphantom{xxx}} = 4,300$

12) $\underline{\hphantom{xxx}} + 670 = 2,230$

13) David sells gems. He finds a diamond in Istanbul and buys it for $3,433. Then, he flies to Cairo and purchases a bigger diamond for the bargain price of $5,922. How much does David spend on the two diamonds?

Subtracting 4–Digit Numbers

Helpful *Hints*	1– Line up the numbers. 2– Start with the units place. (ones place) 3– Regroup if necessary. 4– Subtract the tens place. 5– Continue with other digits.	Example: $5,397$ $- 2,416$ $\overline{2,981}$

✎ Subtract.

1) $8,519 - 5,422$

2) $6,222 - 4,331$

3) $7,821 - 3,212$

4) $8,756 - 6,712$

5) $9,290 - 3,829$

6) $5,117 - 4,216$

✎ Find the missing number.

7) $2223 - \underline{\quad} = 1120$

8) $3574 - \underline{\quad} = 2245$

9) $1124 - 578 = \underline{\quad}$

10) $2300 - \underline{\quad} = 1250$

11) $3780 - 1890 = \underline{\quad}$

12) $2880 - 2560 = \underline{\quad}$

13) Jackson had $3,963 invested in the stock market until he lost $2,171 on those investments. How much money does he have in the stock market now?

Answers of Worksheets – Chapter 2

Adding two–digit numbers

1) 68
2) 46
3) 61
4) 24
5) 73
6) 49
7) 96
8) 75
9) 100

Subtracting two–digit numbers

1) 17
2) 28
3) 50
4) 8
5) 43
6) 69
7) 57
8) 51
9) 42

Adding three–digit numbers

1) 290
2) 657
3) 512
4) 580
5) 862
6) 385
7) 846
8) 1,203
9) 1,045

Adding hundreds

1) 400
2) 500
3) 700
4) 1,000
5) 800
6) 600
7) 1,000
8) 900
9) 800
10) 400
11) 1,000
12) 800
13) 700
14) 800
15) 1,100

Adding 4–digit numbers

1) 7,845
2) 6,486
3) 8,468
4) 5,801
5) 9,451
6) 8,404
7) 131
8) 1,500
9) 1,100
10) 1,300
11) 530
12) 1,560
13) $9,355

Subtracting 4–digit numbers

1) 3,097
2) 1,891
3) 4,609
4) 2,044
5) 5,461
6) 901
7) 1,103
8) 1,329
9) 546
10) 1,050
11) 1,890
12) 320
13) 1,792

Chapter 3: Multiplication and Division

Topics that you'll learn in this chapter:

- ✓ Multiplication
- ✓ Division
- ✓ Long Division by One Digit
- ✓ Division with Remainders

Multiplication

Helpful *Hints*	– Learn the times tables first! – For multiplication, line up the numbers you are multiplying. – Start with the ones place. – Continue with other digits

Example:

$200 \times 90 = 18,000$

✎*Find the answers.*

1)
$$\begin{array}{r} 45 \\ \times\ 13 \\ \hline \\ \hline \end{array}$$

4)
$$\begin{array}{r} 563 \\ \times\ 4 \\ \hline \\ \hline \end{array}$$

7)
$$\begin{array}{r} 478 \\ \times\ 34 \\ \hline \\ \hline \end{array}$$

2)
$$\begin{array}{r} 32 \\ \times\ 10 \\ \hline \\ \hline \end{array}$$

5)
$$\begin{array}{r} 365 \\ \times\ 5 \\ \hline \\ \hline \end{array}$$

8)
$$\begin{array}{r} 956 \\ \times\ 26 \\ \hline \\ \hline \end{array}$$

3)
$$\begin{array}{r} 120 \\ \times\ 9 \\ \hline \\ \hline \end{array}$$

6)
$$\begin{array}{r} 89 \\ \times\ 25 \\ \hline \\ \hline \end{array}$$

9)
$$\begin{array}{r} 391 \\ \times\ 78 \\ \hline \\ \hline \end{array}$$

10) The Haunted House Ride runs 5 times a day. It has 6 cars, each of which can hold 4 people. How many people can ride the Haunted House Ride in one day?

11) Each train car has 3 rows of seats. There are 4 seats in each row. How many seats are there in 5 train cars?

Division

Helpful *Hints*	A typical division problem: Dividend ÷ Divisor = Quotient – In division, we want to find how many times a number (divisor) is contained in another number (dividend). – The result in a division problem is the quotient.

✍ *Find each missing number.*

1) $10 \div \_\_\_ = 1$

2) $48 \div 12 = \_\_\_$

3) $99 \div \_\_\_ = 9$

4) $70 \div 10 = \_\_\_$

5) $44 \div \_\_\_ = 4$

6) $24 \div \_\_\_ = 2$

7) $\_\_\_ \div 10 = 4$

8) $110 \div 11 = \_\_\_$

9) $12 \div \_\_\_ = 1$

10) $90 \div \_\_\_ = 9$

11) $\_\_\_ \div 11 = 8$

12) $\_\_\_ \div 12 = 11$

13) $60 \div \_\_\_ = 6$

14) $\_\_\_ \div 11 = 12$

15) $84 \div 12 = \_\_\_$

16) $80 \div 10 = \_\_\_$

17) $11 \div 11 = \_\_\_$

18) $144 \div \_\_\_ = 12$

19) Anna has 120 books. She wants to put them in equal numbers on 12 bookshelves. How many books can she put on a bookshelf? _____ books

20) If dividend is 99 and the quotient is 11, then what is the divisor? _____

Long Division by One Digit

Helpful	- Remember that long division moves from left to right. - Begin with the number in the left side. (If the dividend is a 3-digit numbers, begin with hundreds place.)
Hints	- If the first digit of dividend is smaller than the divisor, choose another digit from dividend.

✍ *Find the quotient.*

1) $6\overline{)792}$

2) $5\overline{)350}$

3) $6\overline{)174}$

4) $8\overline{)104}$

5) $3\overline{)102}$

6) $9\overline{)189}$

7) $5\overline{)115}$

8) $2\overline{)120}$

9) $7\overline{)112}$

10) $4\overline{)148}$

11) $9\overline{)126}$

12) $6\overline{)240}$

13) $4\overline{)576}$

14) $4\overline{)512}$

15) $9\overline{)1278}$

16) $8\overline{)2768}$

17) $6\overline{)1224}$

18) $4\overline{)3412}$

Division with Remainders

Helpful	- Set up the division problem with the long division bracket.
	- Divide the first digit of the dividend by the divisor.
	- Write the answer on top of the division bracket.
Hints	- Continue with other digits of the dividend.
	- Continue the process until the remainder is smaller than the divisor.

✎ *Find the quotient with remainder.*

1) $5\overline{)592}$

2) $3\overline{)295}$

3) $6\overline{)553}$

4) $5\overline{)214}$

5) $3\overline{)440}$

6) $7\overline{)673}$

7) $4\overline{)213}$

8) $2\overline{)820}$

9) $5\overline{)496}$

10) $6\overline{)791}$

11) $4\overline{)647}$

12) $7\overline{)780}$

13) $4\overline{)5910}$

14) $8\overline{)3515}$

15) $7\overline{)2355}$

16) $9\overline{)1232}$

17) $8\overline{)6029}$

18) $4\overline{)6743}$

Answers of Worksheets – Chapter 3

Multiplication

1) 585
2) 320
3) 1,080
4) 2,252

5) 1,825
6) 2,225
7) 16,252
8) 24,856

9) 30,498
10) 120
11) 60

Division

1) 10
2) 4
3) 11
4) 7
5) 11
6) 12
7) 40

8) 10
9) 12
10) 10
11) 88
12) 132
13) 10
14) 132

15) 7
16) 8
17) 1
18) 12
19) 10
20) 9

Long Division by One Digit

1) 132
2) 70
3) 29
4) 13
5) 34
6) 21

7) 23
8) 60
9) 16
10) 37
11) 14
12) 40

13) 144
14) 128
15) 142
16) 346
17) 204
18) 853

Division with Remainders

1) 118 R4
2) 98 R1
3) 92 R1
4) 42 R4
5) 146 R2
6) 96 R1

7) 53 R1
8) 410 R0
9) 99 R1
10) 131 R5
11) 161 R3
12) 111 R3

13) 1477 R2
14) 439 R3
15) 336 R3
16) 135 R8
17) 753 R5
18) 1685 R3

Chapter 4: Mixed operations

Topics that you'll learn in this chapter:

- ✓ Rounding and Estimating
- ✓ Estimate Sums
- ✓ Estimate Differences
- ✓ Estimate Products
- ✓ Missing Numbers

Rounding and Estimating

Helpful *Hints*	− Rounding is putting a number up or down to the nearest whole number or the nearest hundred, etc. − To estimate means to make a rough guess or calculation. − To round means to simplify a number by scaling it slightly up or down.	**Example:** 64 rounded to the nearest ten is 60, because 64 is closer to 60 than to 70. Estimate: $73 + 69 \approx 70 + 70 = 140$

✎ *Round each number to the underlined place value.*

1) 9̲72

2) 2,9̲95

3) 3̲6̲4

4) 8̲1

5) 5̲5

6) 33̲4

7) 1,2̲03

8) 9.5̲7

9) 7.48̲4

✎ *Estimate the sum by rounding each added to the nearest ten.*

10) 55 + 9

11) 13 + 74

12) 83 + 7

13) 32 + 37

14) 13 + 74

15) 34 + 11

16) 39 + 77

17) 25 + 4

18) 61 + 73

19) 64 + 59

20) 14 + 68

21) 82 + 12

22) 43 + 66

23) 45 + 65

24) 553 + 232

Estimate Sums

Helpful	– To estimate means to make a rough guess or calculation. – For addition, you can round numbers and then solve the problems.	**Example:** 54 + 26 = 50 + 30 = 80
Hints		

✏️ *Estimate the sum by rounding each added to the nearest ten.*

1) 55 + 9

2) 13 + 74

3) 83 + 7

4) 32 + 37

5) 13 + 74

6) 34 + 11

7) 39 + 77

8) 25 + 4

9) 61 + 73

10) 64 + 59

11) 14 + 68

12) 82 + 12

13) 43 + 66

14) 45 + 65

15) 553 + 232

16) 52 + 67

17) 96 + 94

18) 29 + 89

19) 78 + 74

20) 39 + 27

21) 91 + 68

22) 48 + 81

23) 14 + 96

24) 52 + 59

Estimate Differences

Helpful *Hints*	– To estimate means to make a rough guess or calculation. – For subtraction, you can round numbers and then solve the problems.	**Example:** $37 - 12 = 40 - 10 = 30$

✎*Estimate the difference by rounding each number to the nearest ten.*

1) $46 - 11$

2) $23 - 14$

3) $68 - 36$

4) $22 - 13$

5) $59 - 36$

6) $34 - 11$

7) $67 - 37$

8) $38 - 19$

9) $84 - 38$

10) $68 - 48$

11) $58 - 16$

12) $72 - 27$

13) $63 - 33$

14) $49 - 32$

15) $94 - 63$

16) $55 - 32$

17) $87 - 74$

18) $32 - 11$

19) $46 - 39$

20) $99 - 36$

21) $94 - 78$

22) $75 - 23$

23) $99 - 19$

24) $86 - 43$

Estimate Products

Helpful *Hints*	– To estimate means to make a rough guess or calculation. – You can round numbers to estimate the result.

Example:
44 × 17 = 40 × 20 = 800

🖋 *Estimate the products.*

1) 27 × 18

2) 13 × 17

3) 22 × 25

4) 43 × 19

5) 68 × 23

6) 36 × 91

7) 53 × 92

8) 18 × 38

9) 21 × 14

10) 83 × 42

11) 51 × 32

12) 68 × 12

13) 47 × 23

14) 71 × 58

15) 54 × 89

16) 37 × 72

17) 36 × 93

18) 32 × 29

19) 41 × 37

20) 54 × 93

21) 89 × 72

22) 77 × 22

23) 53 × 13

24) 98 × 63

Missing Numbers

Helpful *Hints*	- To find the missing factors in multiplication, you can sometimes use division rule. - Remember, multiplication and division are opposite operations!	Example: ___ × 5 = 45 To solve this problem, simply divide 45 by 5. The answer is 9.

✎ *Find the missing numbers.*

1) $20 \times$ ___ $= 60$

2) $16 \times$ ___ $= 32$

3) ___ $\times 14 = 84$

4) $16 \times$ ___ $= 80$

5) ___ $\times 19 = 38$

6) $17 \times$ ___ $= 34$

7) ___ $\times 1 = 18$

8) $21 \times$ ___ $= 42$

9) $20 \times$ ___ $= 80$

10) $15 \times 7 =$ ___

11) $18 \times 9 =$ ___

12) $21 \times 4 =$ ___

13) $23 \times 7 =$ ___

14) ___ $\times 25 = 75$

15) $24 \times$ ___ $= 120$

16) $22 \times 4 =$ ___

17) $20 \times$ ___ $= 140$

18) $17 \times$ ___ $= 153$

19) ___ $\times 15 = 120$

20) $21 \times 6 =$ ___

21) ___ $\times 22 = 154$

22) $19 \times$ ___ $= 76$

23) $23 \times 9 =$ ___

24) $25 \times 6 =$ ___

25) ___ $\times 18 = 36$

26) $24 \times$ ___ $= 48$

Answers of Worksheets – Chapter 4

Rounding and Estimating

1) 1,000
2) 3,000
3) 360
4) 80
5) 60
6) 330
7) 1,200
8) 9.6

9) 7.5
10) 70
11) 80
12) 90
13) 70
14) 80
15) 40
16) 120

17) 30
18) 130
19) 120
20) 80
21) 90
22) 110
23) 120
24) 780

Estimate sums

1) 70
2) 80
3) 90
4) 70
5) 80
6) 40
7) 120
8) 30

9) 130
10) 120
11) 80
12) 90
13) 110
14) 120
15) 780
16) 120

17) 190
18) 120
19) 150
20) 70
21) 160
22) 130
23) 110
24) 110

Estimate differences

1) 40
2) 10
3) 30
4) 10
5) 20
6) 20
7) 30
8) 20

9) 40
10) 20
11) 40
12) 40
13) 30
14) 20
15) 30
16) 30

17) 20
18) 20
19) 10
20) 60
21) 10
22) 60
23) 80
24) 50

Estimate products

1) 600	9) 200	17) 3600
2) 200	10) 3200	18) 900
3) 600	11) 1500	19) 1600
4) 800	12) 700	20) 4500
5) 1400	13) 1000	21) 6300
6) 3600	14) 4200	22) 1600
7) 4500	15) 4500	23) 500
8) 800	16) 2800	24) 6000

Missing Numbers

1) 3	10) 105	19) 8
2) 2	11) 162	20) 126
3) 6	12) 84	21) 7
4) 5	13) 161	22) 4
5) 2	14) 3	23) 207
6) 2	15) 5	24) 150
7) 18	16) 88	25) 2
8) 2	17) 7	26) 2
9) 4	18) 9	

Chapter 5: Algebra

Topics that you'll learn in this chapter:

- ✓ Evaluating Variable
- ✓ Evaluating Two Variables
- ✓ Solve Equations

Evaluating Variable

Helpful *Hints*	– To evaluate one variable expression, find the variable and substitute a number for that variable. – Perform the arithmetic operations.	**Example:** $4x + 8, x = 6$ $4(6) + 8 = 24 + 8 = 32$

✏ *Simplify each algebraic expression.*

1) $9 - x$, $x = 3$

2) $x + 2$, $x = 5$

3) $3x + 7$, $x = 6$

4) $x + (-5)$, $x = -2$

5) $3x + 6$, $x = 4$

6) $4x + 6$, $x = -1$

7) $10 + 2x - 6$, $x = 3$

8) $10 - 3x$, $x = 8$

9) $\dfrac{20}{x} - 3$, $x = 5$

10) $(-3) + \dfrac{x}{4} + 2x$, $x = 16$

11) $(-2) + \dfrac{x}{7}$, $x = 21$

12) $(-\dfrac{14}{x}) - 9 + 4x$, $x = 2$

13) $(-\dfrac{6}{x}) - 9 + 2x$, $x = 3$

14) $(-2) + \dfrac{x}{8}$, $x = 16$

15) $8(5x - 12)$, $x = -2$

Evaluating Two Variables

Helpful *Hints*	To evaluate an algebraic expression, substitute a number for each variable and perform the arithmetic operations.	**Example:** $2x + 4y - 3 + 2,$ $x = 5, y = 3$ $2(5) + 4(3) - 3 + 2$ $= 10$ $+ 12 - 3 + 2$ $= 21$

✎ *Simplify each algebraic expression.*

1) $2x + 4y - 3 + 2,$

   $x = 5, y = 3$

2) $(-\dfrac{12}{x}) + 1 + 5y,$

   $x = 6, y = 8$

3) $(-4)(-2a - 2b),$

   $a = 5, b = 3$

4) $10 + 3x + 7 - 2y,$

   $x = 7, y = 6$

5) $9x + 2 - 4y,$

   $x = 7, y = 5$

6) $6 + 3(-2x - 3y),$

   $x = 9, y = 7$

7) $12x + y,$

   $x = 4, y = 8$

8) $x \times 4 \div y,$

   $x = 3, y = 2$

9) $2x + 14 + 4y,$

   $x = 6, y = 8$

10) $4a - (5 - b),$

   $a = 4, b = 6$

Solve Equations

Helpful	- The values of two expressions on both sides of an equation are equal.	**Example:**

Helpful

Hints

- The values of two expressions on both sides of an equation are equal.

$$ax + b = c$$

- You only need to perform one Math operation in order to solve the equation.

Example:

$$-8x = 16$$

$$x = -2$$

✎ *Solve each equation.*

1) $x + 3 = 17$

2) $22 = (-8) + x$

3) $3x = (-30)$

4) $(-36) = (-6x)$

5) $(-6) = 4 + x$

6) $2 + x = (-2)$

7) $20x = (-220)$

8) $18 = x + 5$

9) $(-23) + x = (-19)$

10) $5x = (-45)$

11) $x - 12 = (-25)$

12) $x - 3 = (-12)$

13) $(-35) = x - 27$

14) $8 = 2x$

15) $(-6x) = 36$

16) $(-55) = (-5x)$

17) $x - 30 = 20$

18) $8x = 32$

19) $36 = (-4x)$

20) $4x = 68$

21) $30x = 300$

Answers of Worksheets – Chapter 5

Evaluating Variable

1) 6	6) 2	11) 1
2) 7	7) 10	12) −8
3) 25	8) −14	13) −5
4) −7	9) 1	14) 0
5) 18	10) 33	15) −176

Evaluating Two Variables

1) 21	5) 45	9) 58
2) 39	6) −111	10) 17
3) 64	7) 56	
4) 26	8) 6	

Solve Equations

1) 14	8) 13	15) − 6
2) 30	9) 4	16) 11
3) − 10	10) − 9	17) 50
4) 6	11) − 13	18) 4
5) − 10	12) − 9	19) − 9
6) − 4	13) − 8	20) 17
7) − 11	14) 4	21) 10

Chapter 6: Data and Graphs

Topics that you'll learn in this chapter:

- ✓ Graph Points on a Coordinate Plane
- ✓ Bar Graph
- ✓ Tally and Pictographs
- ✓ Line Graphs
- ✓ Stem–And–Leaf Plot
- ✓ Scatter Plots

Graph Points on a Coordinate Plane

Helpful

Hints

- Understand the axes of the coordinate plane:
x-axis goes left and right and y-axis goes up and down.
- You should graph the point in (x, y) form.
- Start at $(0, 0)$.
- Move over x unit to the right if x is positive and move to left if it's negative.
- Move over y units up (if it is positive) or down (if it is negative).

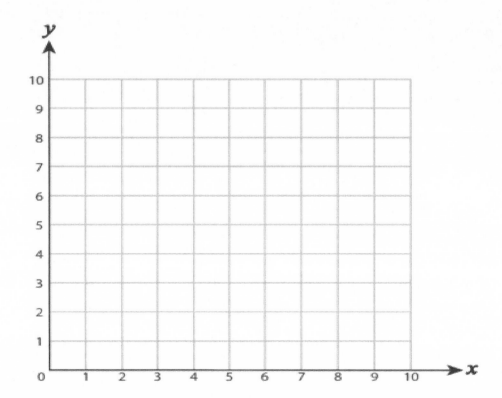

✎ *Plot each point on the coordinate grid.*

1) A (3, 6)   3) C (3, 7)   5) E (5, 2)

2) B (1, 3)   4) D (8, 6)   6) F (9, 3)

Bar Graph

Helpful *Hints*	– A bar graph is a chart that presents data with bars in different heights to match with the values of the data. The bars can be graphed horizontally or vertically.

✍ *Graph the given information as a bar graph.*

Day	Hot dogs sold
Monday	90
Tuesday	70
Wednesday	30
Thursday	20
Friday	60

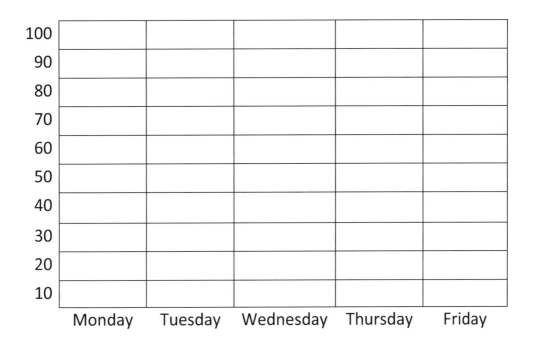

Tally and Pictographs

✎ **Using the key, draw the pictograph to show the information.**

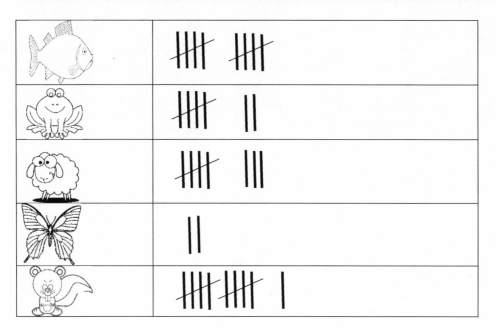

 Key: = 2 animals

Line Graphs

Helpful *Hints*	- Line graphs represent how something changes over time. - There are two arises in a line graph, x-axis is the horizontal line and y-axis is vertical. - x-axis is usually used to show time period and the y-axis shows numbers for what is being represented.

✍ **David work as a salesman in a store. He records the number of shoes sold in five days on a line graph. Use the graph to answer the questions.**

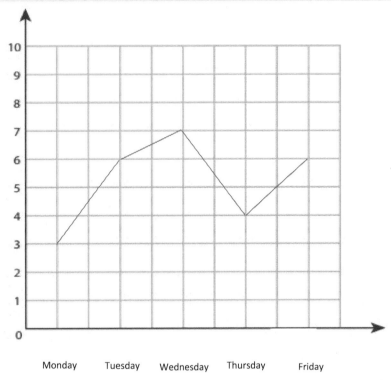

Monday   Tuesday   Wednesday   Thursday   Friday

1) How many cars were sold on Monday?

2) Which day had the minimum sales of shoes?

3) Which day had the maximum number of shoes sold?

4) How many shoes were sold in 5 days?

Stem–And–Leaf Plot

Helpful	— Stem–and–leaf plots display the frequency of the values in a data set.
Hints	— We can make a frequency distribution table for the values, or we can use a stem–and–leaf plot.

Example:

56, 58, 42, 48, 66, 64, 53, 69, 45, 72

Stem	leaf		
4	2	5	8
5	3	6	8
6	4	6	9
7	2		

✍ *Make stem ad leaf plots for the given data.*

1) 74, 88, 97, 72, 79, 86, 95, 79, 83, 91

Stem | Leaf plot

2) 37, 48, 26, 33, 49, 26, 19, 26, 48

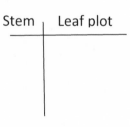

Stem | Leaf plot

3) 58, 41, 42, 67, 54, 65, 65, 54, 69, 53

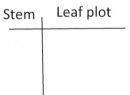

Stem | Leaf plot

Scatter Plots

Helpful	A Scatter (xy) Plot shows the values with points that represent the relationship between two sets of data.
Hints	– The horizontal values are usually x and vertical data is y.

✎ *Construct a scatter plot.*

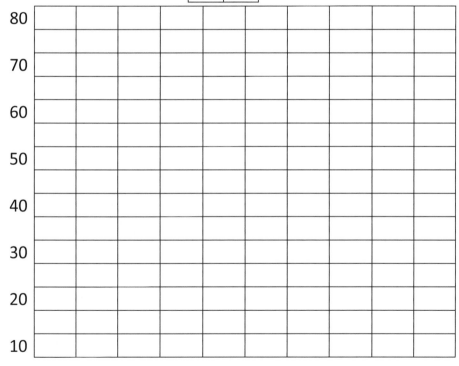

X	Y
1	20
2	40
3	50
4	60

Answers of Worksheets – Chapter 6

Graph Points on a Coordinate Plane

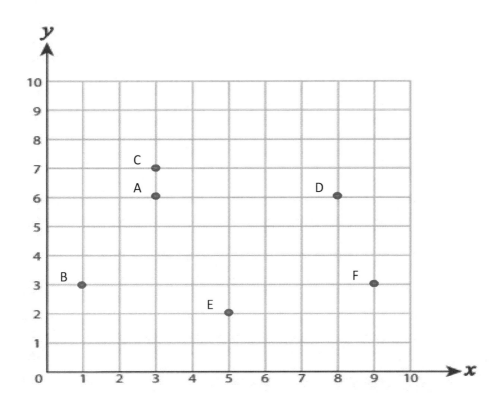

Bar Graph

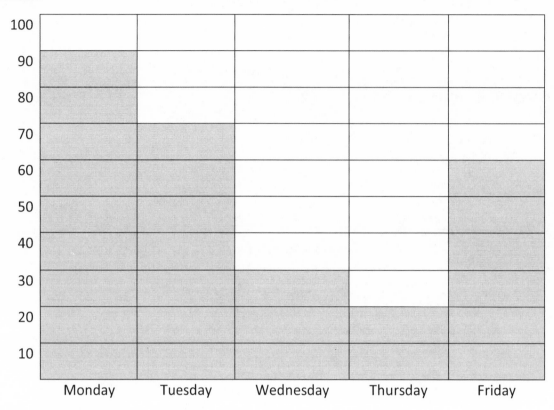

Tally and Pictographs

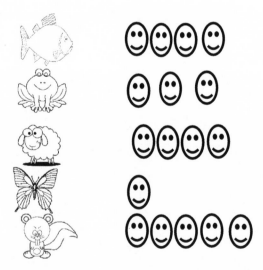

Line Graphs

1)   3

2)   Thursday

3)   Wednesday   4)   26

Stem–And–Leaf Plot

1)

Stem	leaf
7	2 4 9 9
8	3 6 8
9	1 5 7

2)

Stem	leaf
1	9
2	6 6 6
3	3 7
4	8 8 9

3)

Stem	leaf
4	1 2
5	3 4 4 8
6	5 5 7 9

Scatter Plots

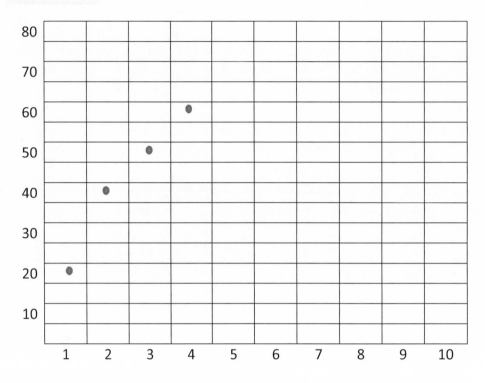

Chapter 7: Patterns and Sequences

Topics that you'll learn in this chapter:

- ✓ Repeating pattern
- ✓ Growing Patterns
- ✓ Patterns: Numbers
- ✓ Patterns

www.EffortlessMath.com

57

Repeating Pattern

Helpful	- Look for a relationship between two shapes in a row.	**Example:**
Hints	- After finding the rule, check the pattern for all other shapes.	
	- Use the rule to find next shapes.	

✎ *Circle the picture that comes next in each picture pattern.*

1)

2)

3)

4)

Growing Patterns

Helpful *Hints*	- Look for a relationship between two shapes in a row. - After finding the rule, check the pattern for all other shapes. - Use the rule to find next shapes.

✎ *Draw the picture that comes next in each growing pattern.*

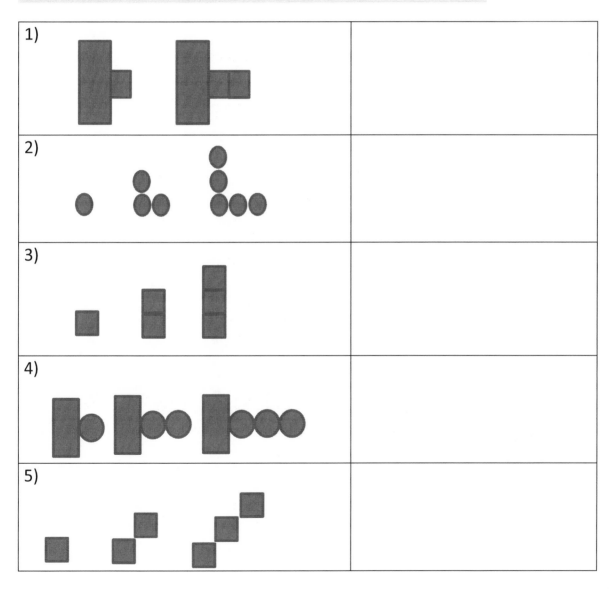

Patterns: Numbers

Helpful	- Look for a relationship between two numbers in a row.	**Example:**
		2, 5, 8, 11, 14, ...
Hints	- After finding the rule, check the pattern for all other numbers.	Rule: The number + 3
	- Use the rule to find next numbers.	2, 5, 8, 11, 14, 17, 20, 23

✍ *Write the numbers that come next.*

1) 3, 6, 9 , 12,   _____, _____, _____, _____

2) 2, 4, 6, 8,   _____, _____, _____, _____

3) 5, 10, 15, 20,   _____, _____, _____, _____

4) 15, 25, 35, 45,   _____, _____, _____, _____

5) 11, 22, 33, 44,   _____, _____, _____, _____

6) 10, 18, 26, 34, 42,   _____, _____, _____, _____

7) 61, 55, 49, 43, 37,   _____, _____, _____, _____

8) 45, 56, 67, 78,   _____, _____, _____, _____

9) 3, 6, 9 , 12, 15, 18, 21, 24   _____, _____, _____, _____

Patterns

Helpful Hints	- A pattern is a group of numbers or objects that follow a rule. - There is a rule in any pattern for repeating and changing.

✎ **Write the next three numbers in each counting sequence.**

1) −32, −23, −14, _____, _____, _____, _____

2) 543, 528, 513, _____, _____, _____, _____

3) _____, _____, 56, 64, _____, 80

4) 23, 34, _____, _____, 67, _____

5) 24, 31, _____, _____, _____

6) 52, 45, _____ , _____, _____

7) 51, 44, 37, _____ , _____ , _____

8) 64, 51, 38, _____ , _____ , _____

9) What are the next three numbers in this counting sequence?

1350, 1550, 1750, _____, _____, _____

10) What is the seventh number in this counting sequence?

7, 16, 25, _____

Answers of Worksheets – Chapter 7

Repeating pattern

Growing patterns

1)

2)

3)

4)

1)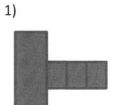

2)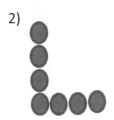

3)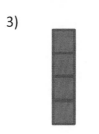

4)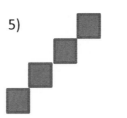

5)

Patterns: Numbers

1) 3, 6, 9, 12, 15, 18, 21, 24
2) 2, 4, 6, 8, 10, 12, 14, 16
3) 5, 10, 15, 20, 25, 30, 35, 40
4) 15, 25, 35, 45, 55, 65, 75, 85
5) 11, 22, 33, 44, 55, 66, 77, 88
6) 10, 18, 26, 34, 42, 50, 58, 66
7) 61, 55, 49, 43, 37, 31, 25, 19
8) 45, 56, 67, 78, 89, 100, 111, 122

Patterns

1) −5, 4, 13, 22
2) 498, 483, 468
3) 40−48−56−64−72−80
4) 23−34−45−56−67−78
5) 38−45−52
6) 38−31−24
7) 30,23,16
8) 25,12, −1
9) 1950−2150−2350
10) 61

Chapter 8: Money

Topics that you'll learn in this chapter:

- ✓ Add Money Amounts
- ✓ Subtract Money Amounts
- ✓ Money: Word Problems

Add Money Amounts

Helpful	Example:
Hints	$12.23 +$26.55 $38.78

✎ *Subtract.*

1)
$314
+$152
_____

$624
+$410
_____

$390
+$215
_____

2)
$321
+$430
_____

$530
+$321
_____

$712
+$145
_____

3)
$411
+$316
_____

$559
+$228
_____

$731
+$213
_____

4)
$621
+$168
_____

$321
+$129
_____

$615
+$371
_____

Subtract Money Amounts

Helpful	- For subtracting money, start with the cents and then the dollars.	**Example:**
Hints	- Move from right to left. - Borrow if it's necessary!	$87.47 −$26.44 $61.03

Subtract.

5)
$825
−$166
_____

$651
−$110
_____

$754
−$565
_____

6)
$539
−$137
_____

$498
−$359
_____

$992
−$549
_____

7)
$436
−$219
_____

$512
−$128
_____

$632
−$444
_____

8) Linda had $12.00. She bought some game tickets for $7.14. How much did she have left?

Money: Word Problems

Helpful *Hints*	- **First review key words in word problems and what they mean.** **Addition:** total, in all, together, and, more than, sum, etc. **Subtraction:** difference, less than, minus, left over, etc. **Multiplication:** times, product, each, double, etc. **Division:** split, row, half, fourth, etc. **Example:** Peter has 211 blue marbles and 29 red marbles. How many marbles does he have **in all**? (Add 211 + 29 = 240 cards).

✏️ *Solve.*

1) How many boxes of envelopes can you buy with $18 if one box costs $3?

2) After paying $6.25 for a salad, Ella has $45.56. How much money did she have before buying the salad?

3) How many packages of diapers can you buy with $50 if one package costs $5?

4) Last week James ran 20 miles more than Michael. James ran 56 miles. How many miles did Michael run?

5) Last Friday Jacob had $32.52. Over the weekend he received some money for cleaning the attic. He now has $44. How much money did he receive?

6) After paying $10.12 for a sandwich, Amelia has $35.50. How much money did she have before buying the sandwich?

Answers of Worksheets – Chapter 8

Add Money Amounts

1) 466, 1,034, 605
2) 751, 851, 857
3) 727, 787, 944
4) 789, 450, 986

Subtract Money Amounts

1) 659–541–189
2) 402–139–443
3) 217–384–188
4) 4.86

Money: word problem

1) 6
2) $51.81
3) 10
4) 36
5) 11.48
6) 45.62

Chapter 9: Measurement

Topics that you'll learn in this chapter:

- ✓ Metric System
- ✓ Length
- ✓ Temperature
- ✓ Liters & Milliliters
- ✓ Kilograms & Grams

Metric System

Helpful *Hints*	1 m = 100 cm 1 cm = 10 mm 1m = 1000 mm 1 km = 1000 m	**Example:** 12 cm = 0.12 m

✎ **Convert to the units.**

1) 4 mm = _____ cm

2) 0.6 m = _____ mm

3) 2 m = _____ cm

4) 0.03 km = _____ m

5) 3000 mm = _____ km

6) 5 cm _____ m

7) 0.03 m = _____ cm

8) 1000 mm = _____ km

9) 600 mm = _____ m

10) 0.77 km = _____ mm

11) 0.08 km = _____ m

12) 0.30 m = _____ cm

13) 400 m = _____ km

14) 5000 cm = _____ km

15) 40 mm = _____ cm

16) 800 m = _____ km

Length

| *Helpful* | 1 foot = 12 inches | **Example:** |
| *Hints* | 1 yard = 3 feet
1meter = 100 centimeters
1 kilometer = 1000 meters | 3 meter = 300 cm |

✎**Use a ruler to find the length of the line segment below to the nearest quarter inch.**

✎ **Convert the following measurements.**

1) 2 feet = _____ inches

2) 5 feet = _____ inches

3) 1 yard = _____ feet

4) 3 yards = _____ feet

5) 1 meter = _____ centimeter

6) 3 kilometers = _____ meters

7) 100 meters = _____ centimeters

8) 8 yards = _____ feet

Temperature

Helpful	- **Convert Celsius to Fahrenheit:** Multiply the temperature by 1.8 and then add 32 to the result	**Example:** $32 \,°F = 0 \,°C$
Hints	- **Convert Fahrenheit to Celsius:** Subtract 32 from the number and then divide the result by 1.8	$10.00 \,°C = 50 \,°F$

1) ✍ *What temperature is shown on this Celsius thermometer?*

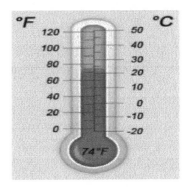

✍ *Convert Celsius into Fahrenheit.*

2) $10°C =$ ___ $°F$ 6) $25°C =$ ___ $°F$ 11) $30°C =$ ___ $°F$

3) $35°C =$ ___ $°F$ 8) $50°C =$ ___ $°F$ 12) $20°C =$ ___ $°F$

4) $80°C =$ ___ $°F$ 9) $45°C =$ ___ $°F$ 7) $70°C =$ ___ $°F$

5) $15°C =$ ___ $°F$ 10) $90°C =$ ___ $°F$ 13) $80°C =$ ___ $°F$

Liters & Milliliters

Helpful	1 L
	=
Hints	1000 ml

✎*Solve.*

1) 10 l = _____ ml

2) 4 l = _____ ml

3) 20 l = _____ ml

4) 24 l = _____ ml

5) 27 l = _____ ml

6) 14 l = _____ ml

7) 50 l = _____ ml

8) 45 l = _____ ml

9) 98 l = _____ ml

10) 1000 ml = _____ l

11) 3000 ml = _____ l

12) 70, 000 ml = _____ l

13) 6000 ml = _____ l

14) 13, 000 ml = _____ l

15) 8000 ml = _____ l

16) 30, 000 ml = _____ l

17) 9000 ml = _____ l

18) 10, 000 ml = _____ l

Kilograms & Grams

Helpful	1 kg
Hints	=
	1000 g

✍ *Solve.*

1) 10 kg = _____ g

2) 33 kg = _____ g

3) 100 kg = _____ g

4) 60 kg = _____ g

5) 85 kg = _____ g

6) 120 kg = _____ g

7) 28 kg = _____ g

8) 72 kg = _____ g

9) 56 kg = _____ g

10) 100,000 g = _____ kg

11) 30, 000 g = _____ kg

12) 70, 000 g = _____ kg

13) 600, 000 g = _____ kg

14) 130, 000 g = _____ kg

15) 80,000 g = _____ kg

16) 300, 000 g = _____ kg

17) 90, 000 g = _____ kg

18) 10, 000 g = _____ kg

Answers of Worksheets – Chapter 9

Metric System

1) 4 mm = 0.4 cm
2) 0.6 m = 600 mm
3) 2 m = 200 cm
4) 0.03 km = 30 m
5) 3000 mm = 0.003 km
6) 5 cm = 0.05 m
7) 0.03 m = 3 cm
8) 1000 mm = 0.001 km
9) 600 mm = 0.6 m
10) 0.77 km = 770,000 mm
11) 0.08 km = 80 m
12) 0.30 m = 30 cm
13) 400 m = 0.4 km
14) 5000 cm = 0.05 km
15) 40 mm = 4 cm
16) 800 m = 0.8 km

Length

1) 24
2) 60
3) 3
4) 9
5) 100
6) 3000
7) 10000
8) 24

Temperature

1) 20
2) 50
3) 95
4) 176
5) 59
6) 77
7) 122
8) 113
9) 194
10) 86
11) 68
12) 158
13) 176

Liters & Milliliters

1) 10 l = 10,000 ml
2) 4 l = 4,000 ml
3) 20 l = 20,000 ml
4) 24 l = 24,000 ml
5) 27 l = 27,000 ml
6) 14 l = 14,000 ml
7) 50 l = 50,000 ml
8) 45 l = 45,000 ml
9) 98 l = 98,000 ml
10) 1,000 ml = 1 L
11) 3,000 ml = 3 L
12) 70,000 ml = 70 L
13) 6,000 ml = 6 L
14) 13,000 ml = 13 L
15) 8,000 ml = 8 L
16) 30,000 ml = 30 L
17) 9,000 ml = 9 L
18) 10,000 ml = 10 L

Kilograms & Grams

1) 10 kg = 10,000 g
2) 33 kg = 33,000 g
3) 100 kg = 100,000 g
4) 60 kg = 60,000 g
5) 85 kg = 85,000 g
6) 120 kg = 120,000g
7) 28 kg = 28,000 g
8) 72 kg = 72,000 g
9) 56 kg = 56,000 g
10) 100,000 g = 100 kg
11) 30,000 g = 30 kg
12) 70, 000 g = 70 kg
13) 600,000 g = 600 kg
14) 130,000 g = 130 kg
15) 80,000 g = 80 kg
16) 300,000 g = 300 kg
17) 90,000 g = 90 kg
18) 10,000 g = 10 kg

Chapter 10: Time

Topics that you'll learn in this chapter:

- ✓ Read Clocks
- ✓ Telling Time
- ✓ Digital Clock
- ✓ Measurement – Time

Read Clocks

Helpful Hints	- A clock has a round face with the numbers 1 through 12 on it. - It has an hour hand and a minute hand to show the hour and minute. - To tell the hour, look at the number that the short hand just passed. - To tell the minute, look at the number the longer hand just passed.

✎ *Write the time below each clock.*

1)

_____

2)

_____

3)

_____

4)

_____

5)

_____

6)

_____

Telling Time

1) What time is shown by this clock?

_____

2) It is night. What time is shown on this clock?

_____

✎How much time has passed?

3) From 1:15 AM to 4:35 AM: _____ hours and _____ minutes.

4) From 1:25 AM to 4:05 AM: _____ hours and _____ minutes.

5) It's 8:30 P.M. What time was 5 hours ago?

_____ O'clock

Digital Clock

Helpful	- A digital clock contains two numbers separated by a colon. The first number marks the hour and the second number, found after the colon, shows the minutes.
Hints	- For example, "3 : 14" means the time is "three-fourteen" or "fourteen past three."

✎ *What time is it? Write the time in words in front of each.*

1) 2 : 30   _____

2) 3 : 15   _____

3) 5 : 45   _____

4) 9 : 20   _____

5) 10 : 5   _____

6) 12 : 50   _____

Measurement – Time

Helpful	60 minutes = 1 hour 60 seconds = 1 minute	**Example:**
Hints		4 hours = 240 minutes

✎ *How much time has passed?*

1) 1:15 AM to 4:35 AM: _____ hours and _____ minutes.

2) 2:35 AM to 5:10 AM: _____ hours and _____ minutes.

3) 6:00 AM. to 7:25 AM. = _____ hour(s) and _____ minutes.

4) 6:15 PM to 7:30 PM. = _____ hour(s) and _____ minutes

5) 5:15 A.M. to 5:45 A.M. = _____ minutes

6) 4:05 A.M. to 4:30 A.M. = _____ minutes

7) There are _____ second in 15 minutes.

8) There are _____ second in 11 minutes.

9) There are _____ second in 27 minutes.

10) There are _____ minutes in 10 hours.

11) There are _____ minutes in 20 hours.

12) There are _____ minutes in 12 hours.

Answers of Worksheets – Chapter 10

Read clocks

1) 1
2) 4 : 45
3) 8
4) 3 : 30
5) 10 : 15
6) 8 : 35

Telling Time

1) 12:00
2) 22:10 PM
3) 3 hours and 20 minutes
4) 2 hours and 40 minutes
5) 3:30 PM

Digital Clock

1) It's two thirty.
2) It's three Fifteen.
3) It's five forty–five.
4) It's nine twenty.
5) It's ten five.
6) It's Twelve Fifty.

Measurement – Time

1) 3:20
2) 2:35
3) 1:25
4) 1:15
5) 30 minutes
6) 25 minutes
7) 900
8) 660
9) 1,620
10) 600
11) 1,200
12) 720

Chapter 11: Geometric

Topics that you'll learn in this chapter:

- ✓ Identifying Angles: Acute, Right, Obtuse, and Straight Angles
- ✓ Polygon Names
- ✓ Classify Triangles
- ✓ Parallel Sides in Quadrilaterals
- ✓ Identify Rectangles
- ✓ Perimeter: Find the Missing Side Lengths
- ✓ Perimeter and Area of Squares
- ✓ Perimeter and Area of rectangles
- ✓ Find the Area or Missing Side Length of a Rectangle
- ✓ Area and Perimeter: Word Problems
- ✓ Area of Squares and Rectangles
- ✓ Volume of Cubes and Rectangle Prisms

Identifying Angles: Acute, Right, Obtuse, and Straight Angles

Helpful

Hints

Angle Type	Measure Range
acute	0° to 90°
right	90°
Obtuse	90° to 180°
straight	180°
reflex	180° to 360°

✎ **Write the name of the angles.**

1)

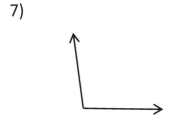

2)

3)

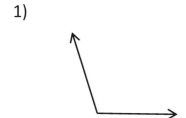

4)

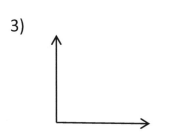

5)

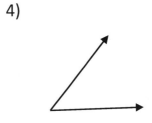

6)

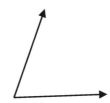

7)

8)

Polygon Names

	Name	Shape	Name	Shape
Helpful	Triangle (or Trigon)	△	Octagon	⬡
Hints	Quadrilateral (or Tetragon)	□	Nonagon (or Enneagon)	⬡
	Pentagon	⬠	Decagon	⬡
	Hexagon	⬡	Hendecagon (or Undecagon)	⬡
	Heptagon	⬡	Dodecagon	⬡

✎ Write name of polygons.

1)

2)

3)

4)

5)

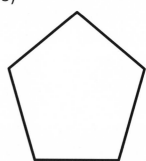

6)

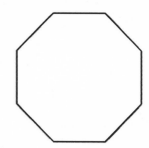

Classify Triangles

✍ *Classify the triangles by their sides and angles.*

1)
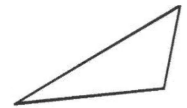

2)
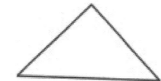

3)

4)
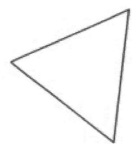

5)

6)
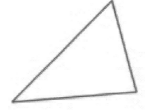

Parallel Sides in Quadrilaterals

Helpful *Hints*	**Quadrilaterals:**	
	Square	Rhombus
	Rectangle	Trapezoid
	Parallelogram	Kike

✎ *Write name of quadrilaterals.*

1)

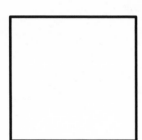

2)

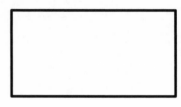

3)

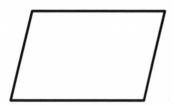

4)

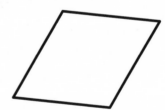

5)

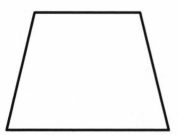

6)

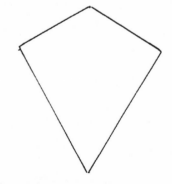

Identify Rectangles

Helpful	– A rectangle is a quadrilateral with two pairs of congruent parallel sides and four right angles.	Example:
Hints		

✏ Solve.

1) A rectangle has _____ sides and _____ angles.

2) Draw a rectangle that is 6 centimeters long and 3 centimeters wide. What is the perimeter?

3) Draw a rectangle 5 cm long and 2 cm wide.

4) Draw a rectangle whose length is 4 cm and whose width is 2 cm. What is the perimeter of the rectangle?

5) What is the perimeter of the rectangle?

8

4

Perimeter: Find the Missing Side Lengths

Helpful	Perimeter of square = 4 × *side*	Perimeter of rectangles = 2(*length* + *width*)

Helpful

Perimeter of square = 4 × *side*

side ▢

Hints

Perimeter of rectangles = 2(*length* + *width*)

width ▭

length

✍️ *Find the missing side of each shape.*

1) perimeter = 44

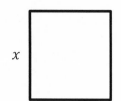

x

2) perimeter = 28

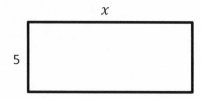

x

5

3) perimeter = 30

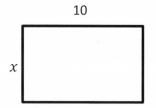

10

x

4) perimeter = 16

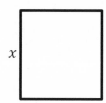

x

5) perimeter = 60

x

6) perimeter = 22

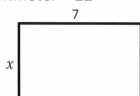

7

x

7) perimeter = 30

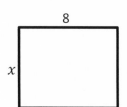

8

x

8) perimeter = 36

x

Perimeter and Area of Squares

Helpful

Hints

Perimeter = $4 \times side$

Area = $(side) \times (side)$

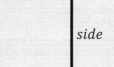

side

✎*Find perimeter and area of squares.*

1) A: _____, P: _____

5

2) A: _____, P: _____

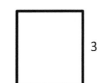

3

3) A: _____, P: _____

7

4) A: _____, P: _____

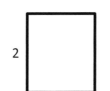

2

5) A: _____, P: _____

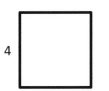

4

6) A: _____, P: _____

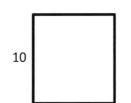

10

7) A: _____, P: _____

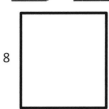

8

8) A: _____, P: _____

12

Perimeter and Area of rectangles

> *Helpful*
>
> *Hints*
>
> P = 2(*length* + width)
>
> A = *length* × *width*
>
> length
>
> width

✏️ *Find perimeter and area of rectangles.*

1) A: _____, P: _____

10

5

2) A: _____, P: _____

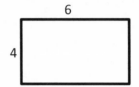

6

4

3) A: _____, P: _____

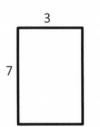

3

7

4) A: _____, P: _____

15

10

5) A: _____, P: _____

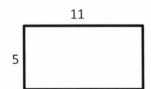

11

5

6) A: _____, P: _____

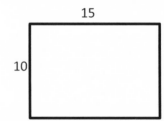

9

8

Find the Area or Missing Side Length of a Rectangle

Helpful

Hints

$A = Width \times Length$

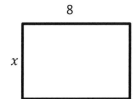

Find area or missing side length of rectangles.

1) Area = ?

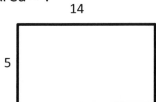

2) Area = 48, x = ?

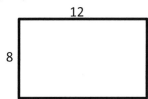

3) Area = 40, x = ?

4) Area = ?

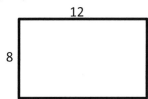

5) Area = ?

6) Area = 600,
   x = ?

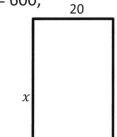

7) Area = 384, x = ?

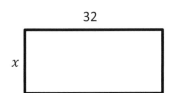

8) Area = 525, x = ?

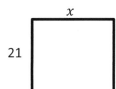

Area and Perimeter: Word Problems

Helpful	- Area of square = (one side of the square)^2 - Area of Rectangle = length × width
Hints	- Perimeter of square = 4 × one side of the square. Perimeter of rectangle = 2 × (length + width)

✎ *Solve.*

1) The area of a rectangle is 72 square meters. The width is 8 meters. What is the length of the rectangle?

2) A square has an area of 36 square feet. What is the perimeter of the square?

3) Ava built a rectangular vegetable garden that is 6 feet long and has an area of 54 square feet. What is the perimeter of Ava's vegetable garden?

4) A square has a perimeter of 64 millimeters. What is the area of the square?

5) The perimeter of David's square backyard is 44 meters. What is the area of David's backyard?

6) The area of a rectangle is 40 square inches. The length is 8 inches. What is the perimeter of the rectangle?

Area of Squares and Rectangles

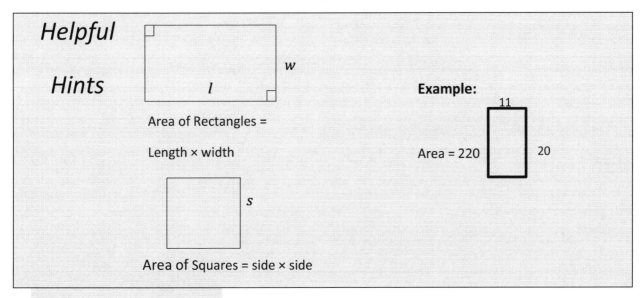

Helpful

Hints

Area of Rectangles =

Length × width

Area of Squares = side × side

Example:

11

Area = 220

20

🖎*Find the area of each.*

1)

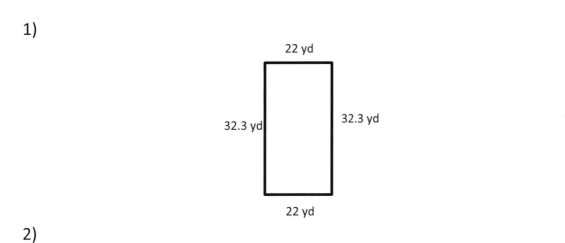

22 yd

32.3 yd

32.3 yd

22 yd

2)

27mi

27 mi

27 mi

27 mi

Volume of Cubes and Rectangle Prisms

Helpful	– Volume is the amount of space inside of a solid figure, like a rectangle prism, cube, or cylinder.
	– Volume of a cube = (one side)^3
Hints	– Volume of a rectangle prism: Length × Width × Height

Find the volume of each of the rectangular prisms.

1)

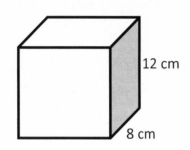

14 cm
12 cm
8 cm

2)

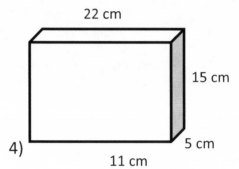

22 cm
15 cm
5 cm
11 cm

3)

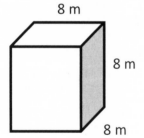

8 m
8 m
8 m

4)

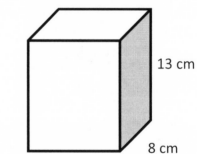

13 cm
8 cm

5)

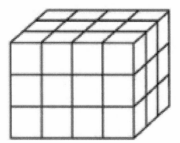

6)

Answers of Worksheets – Chapter 11

Identifying Angles: Acute, Right, Obtuse, and Straight Angles

1) Obtuse
2) Acute
3) Right
4) Acute
5) Straight
6) Obtuse
7) Obtuse
8) Acute

Polygon Names

1) Triangle
2) Quadrilateral
3) Pentagon
4) Hexagon
5) Heptagon
6) Octagon

Classify triangles

1) Scalene, obtuse
2) Isosceles, right
3) Scalene, right
4) Equilateral, acute
5) Scalene, acute
6) Scalene, acute

Parallel Sides in Quadrilaterals

1) Square
2) Rectangle
3) Parallelogram
4) Rhombus
5) Trapezoid
6) Kike

Identify Rectangles

1) 4 - 4
2) 18
3) Use a rule to draw the rectangle
4) 12
5) 24

Perimeter: Find the Missing Side Lengths

1) 11
2) 9
3) 5
4) 4
5) 15
6) 4
7) 7
8) 9

Perimeter and Area of Squares

1) A: 25, P: 20
2) A: 9, P: 12
3) A: 49, P: 28
4) A: 4, P: 8
5) A: 16, P: 16
6) A: 100, P: 40
7) A: 64, P: 32
8) A: 144, P: 48

Perimeter and Area of rectangles

1) A: 50, P: 30
2) A: 24, P: 20
3) A: 21, P: 20
4) A: 150, P: 50
5) A: 55, P: 32
6) A: 72, P: 34

Find the Area or Missing Side Length of a Rectangle

1) 70
2) 6
3) 10
4) 96
5) 330
6) 30
7) 12
8) 25

Area and Perimeter: Word Problems

1) 9
2) 24
3) 30
4) 256
5) 121
6) 26

Area of Squares and Rectangles

1) 710.6 yd^2

2) 729 mi^2

Volume of Cubes and Rectangle Prisms

1) 1344 cm^3

2) 1650 cm^3

3) 512 m^3

4) 1144 cm^3

5) 36
6) 44

Chapter 12: Three-Dimensional Figures

Topics that you'll learn in this chapter:

- ✓ Identify Three–Dimensional Figures
- ✓ Count Vertices, Edges, and Faces
- ✓ Identify Faces of Three–Dimensional Figures

Identify Three–Dimensional Figures

Helpful

Hints

– List of figures in this page:

Square pyramid   Triangular pyramid
Triangular prism   Hexagonal prism
Rectangular prism   Pentagonal prism
Cube

✎ *Write the name of each shape.*

1)

2)

3)

4)

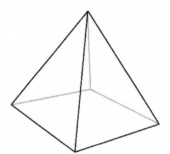

5)

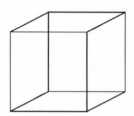

6)

7)

Count Vertices, Edges, and Faces

Helpful	− **Vertex:** the point at which two sides of a polygon meet.
	− **Edge:** It's a line segment joining two vertices in a polygon.
Hints	− **Face:** It is each flat side of a solid.

	Number of edges	Number of faces	Number of vertices
1)	_____	_____	_____
2)	_____	_____	_____
3)	_____	_____	_____
4)	_____	_____	_____
5)	_____	_____	_____
6)	_____	_____	_____

Identify Faces of Three–Dimensional Figures

Helpful *Hints*	– A face is a flat surface. All solid figures, with the exception of a sphere, has one or more faces.

✎ *Write the number of faces.*

1)

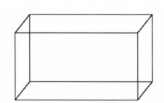

2)

3)

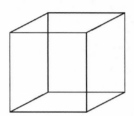

4)

5)

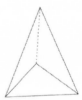

6)

7)

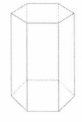

8)

Answers of Worksheets – Chapter 12

Identify Three–Dimensional Figures

1) Cube
2) Triangular pyramid
3) Triangular prism
4) Square pyramid

5) Rectangular prism
6) Pentagonal prism
7) Hexagonal prism

Count Vertices, Edges, and Faces

	Number of edges	Number of faces	Number of vertices
1)	6	4	4
2)	8	5	5
3)	12	6	8
4)	12	6	8

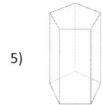

5)   15   7   10

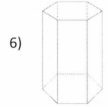

6)   18   8   12

Identify Faces of Three–Dimensional Figures

1) 6
2) 2
3) 5
4) 4
5) 6
6) 7
7) 8
8) 5

Chapter 13: Symmetry and Transformations

Topics that you'll learn in this chapter:

- ✓ Line Segments
- ✓ Identify Lines of Symmetry
- ✓ Count Lines of Symmetry
- ✓ Parallel, Perpendicular and Intersecting Lines

Line Segments

Helpful	– Line segment has a beginning and end points.
	– Line doesn't have beginning or end points.
Hints	– Ray starts out at a point and continues off to infinity.

✍️ *Write each as a line, ray or line segment.*

1)

2)

3)
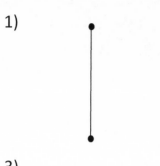

4)

5)

6)

7)

8)

Identify Lines of Symmetry

Helpful	You can find if a shape has a Line of Symmetry by folding it.
Hints	

✎ *Tell whether the line on each shape is a line of symmetry.*

1)

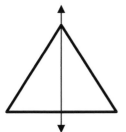

2)

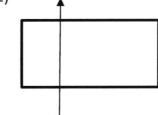

3)

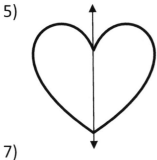

4)

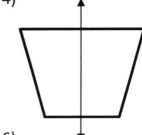

5)

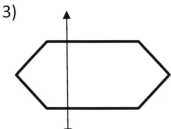

6)

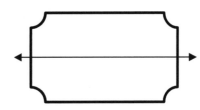

7)

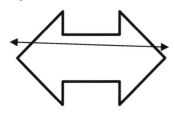

8)

Count Lines of Symmetry

Helpful	– Line of symmetry cuts a shape into two same shapes. – A shape can have many lines of symmetry.
Hints	

✍ *Draw lines of symmetry on each shape. Count and write the lines of symmetry you see.*

1)

2)
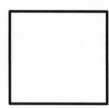

3)
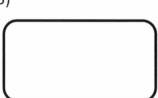

4)

5)

6)
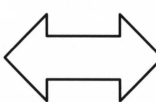

7)
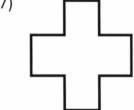

8)
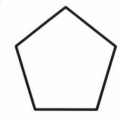

Parallel, Perpendicular and Intersecting Lines

Helpful *Hints*	− Intersecting lines meet each other at a point. − Parallel lines never intersect. They are always equidistant. − Perpendicular lines are two lines that intersect to form right angles.

✎ **State whether the given pair of lines are parallel, perpendicular, or intersecting.**

1)

2)

3)

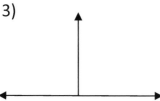

4)

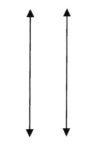

5)

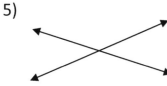

6)

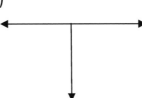

7)

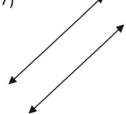

8)

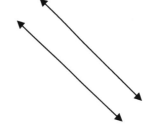

Answers of Worksheets – Chapter 13

Line Segments

1) Line segment
2) Ray
3) Line
4) Line segment

5) Ray
6) Line
7) Line
8) Line segment

Identify lines of symmetry

1) yes
2) no
3) no

4) yes
5) yes
6) yes

7) no
8) yes

Count lines of symmetry

1)

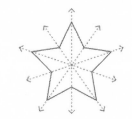

2)

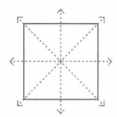

3)

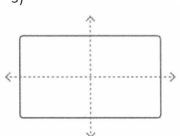

4)

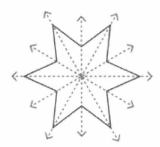

5)

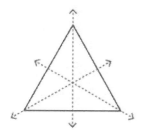

6)

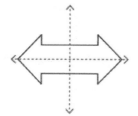

7)

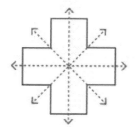

8)

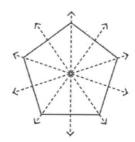

Parallel, Perpendicular and Intersecting Lines

1) Parallel
2) Intersection
3) Perpendicular
4) Parallel
5) Intersection
6) Perpendicular
7) Parallel
8) Parallel

Chapter 14: Fractions

Topics that you'll learn in this chapter:

- ✓ Fractions
- ✓ Add Fractions with Like Denominators
- ✓ Subtract Fractions with Like Denominators
- ✓ Add and Subtract Fractions with Like Denominators
- ✓ Compare Sums and Differences of Fractions with Like Denominators
- ✓ Add 3 or More Fractions with Like Denominators
- ✓ Simplifying Fractions
- ✓ Add Fractions with Unlike Denominators
- ✓ Subtract Fractions with Unlike Denominators
- ✓ Add Fractions with Denominators of 10 and 100
- ✓ Add and Subtract Fractions with Denominators of 10, 100, and 1000

Fractions

Helpful	- We use fractions for measuring. For example, one third, means one part of three parts of something.
Hints	- In fractions, we have two numbers. The number in the bottom shows how many parts are in a whole of something. The top number shows how many parts of something have a feature. For example: $\frac{1}{2}$ means half

✍ What fraction of the squares is shaded?

1)

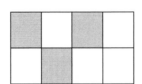

2)

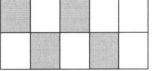

3)

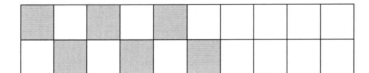

✍ Which fraction has the least value?

4) $\frac{1}{3}$ $\frac{2}{7}$ $\frac{8}{21}$ $\frac{4}{42}$

5) $\frac{1}{2}$ $\frac{3}{8}$ $\frac{3}{4}$ $\frac{9}{16}$

Add Fractions with Like Denominators

Helpful	-	Adding with the same denominator:	**Example:**
Hints		$\dfrac{1}{6} + \dfrac{2}{6} = \dfrac{1+2}{6} = \dfrac{3}{6}$	$\dfrac{2}{5} + \dfrac{1}{5} = \dfrac{2+1}{5} = \dfrac{3}{5}$

✎*Add fractions.*

1) $\dfrac{2}{3} + \dfrac{1}{3}$

2) $\dfrac{3}{5} + \dfrac{2}{5}$

3) $\dfrac{5}{8} + \dfrac{4}{8}$

4) $\dfrac{3}{4} + \dfrac{3}{4}$

5) $\dfrac{4}{10} + \dfrac{3}{10}$

6) $\dfrac{3}{7} + \dfrac{2}{7}$

7) $\dfrac{4}{5} + \dfrac{4}{5}$

8) $\dfrac{5}{14} + \dfrac{7}{14}$

9) $\dfrac{5}{18} + \dfrac{11}{18}$

10) $\dfrac{3}{12} + \dfrac{5}{12}$

11) $\dfrac{5}{13} + \dfrac{5}{13}$

12) $\dfrac{8}{25} + \dfrac{12}{25}$

13) $\dfrac{9}{15} + \dfrac{6}{15}$

14) $\dfrac{4}{20} + \dfrac{5}{20}$

15) $\dfrac{9}{17} + \dfrac{3}{17}$

16) $\dfrac{18}{32} + \dfrac{15}{32}$

17) $\dfrac{12}{28} + \dfrac{10}{28}$

18) $\dfrac{4}{20} + \dfrac{8}{20}$

19) $\dfrac{24}{45} + \dfrac{11}{45}$

20) $\dfrac{8}{36} + \dfrac{18}{36}$

21) $\dfrac{19}{30} + \dfrac{12}{30}$

Subtract Fractions with Like Denominators

Helpful	- Subtracting with the same denominator:	**Example:**
Hints	$\dfrac{5}{6} - \dfrac{1}{6} = \dfrac{5-1}{6} = \dfrac{4}{6}$	$\dfrac{4}{7} - \dfrac{2}{7} = \dfrac{2}{7}$

✍ *Subtract fractions.*

1) $\dfrac{4}{5} - \dfrac{2}{5}$

2) $\dfrac{2}{3} - \dfrac{1}{3}$

3) $\dfrac{7}{9} - \dfrac{4}{9}$

4) $\dfrac{5}{6} - \dfrac{3}{6}$

5) $\dfrac{4}{10} - \dfrac{3}{10}$

6) $\dfrac{5}{7} - \dfrac{3}{7}$

7) $\dfrac{7}{8} - \dfrac{5}{8}$

8) $\dfrac{11}{13} - \dfrac{9}{13}$

9) $\dfrac{8}{10} - \dfrac{5}{10}$

10) $\dfrac{8}{12} - \dfrac{7}{12}$

11) $\dfrac{18}{21} - \dfrac{12}{21}$

12) $\dfrac{15}{19} - \dfrac{9}{19}$

13) $\dfrac{9}{25} - \dfrac{6}{25}$

14) $\dfrac{25}{32} - \dfrac{17}{32}$

15) $\dfrac{22}{27} - \dfrac{9}{27}$

16) $\dfrac{27}{30} - \dfrac{15}{30}$

17) $\dfrac{31}{33} - \dfrac{26}{33}$

18) $\dfrac{18}{28} - \dfrac{8}{28}$

19) $\dfrac{35}{40} - \dfrac{15}{40}$

20) $\dfrac{29}{35} - \dfrac{19}{35}$

21) $\dfrac{21}{36} - \dfrac{11}{36}$

Add and Subtract Fractions with Like Denominators

Helpful	- Adding and Subtracting with the same denominator:	**Example:**
Hints	$$\frac{a}{b} + \frac{c}{b} = \frac{a+c}{b}$$ $$\frac{a}{b} - \frac{c}{b} = \frac{a-c}{b}$$	$$\frac{3}{12} - \frac{1}{12} = \frac{2}{12}$$ $$\frac{7}{9} - \frac{3}{9} = \frac{4}{9}$$

✍ **Add fractions.**

1) $\dfrac{1}{3} + \dfrac{2}{3}$

2) $\dfrac{3}{6} + \dfrac{2}{6}$

3) $\dfrac{5}{8} + \dfrac{2}{8}$

4) $\dfrac{3}{9} + \dfrac{5}{9}$

5) $\dfrac{4}{10} + \dfrac{1}{10}$

6) $\dfrac{3}{7} + \dfrac{2}{7}$

7) $\dfrac{3}{5} + \dfrac{2}{5}$

8) $\dfrac{1}{12} + \dfrac{1}{12}$

9) $\dfrac{16}{25} + \dfrac{5}{25}$

✍ **Subtract fractions.**

10) $\dfrac{4}{5} - \dfrac{2}{5}$

11) $\dfrac{5}{7} - \dfrac{3}{7}$

12) $\dfrac{3}{4} - \dfrac{2}{4}$

13) $\dfrac{8}{9} - \dfrac{3}{9}$

14) $\dfrac{6}{14} - \dfrac{3}{14}$

15) $\dfrac{4}{15} - \dfrac{1}{15}$

16) $\dfrac{15}{16} - \dfrac{13}{16}$

17) $\dfrac{25}{50} - \dfrac{20}{50}$

18) $\dfrac{10}{21} - \dfrac{7}{21}$

Compare Sums and Differences of Fractions with Like Denominators

Helpful *Hints*	- When fractions have same denominators (the bottom number), the one with bigger numerator (the top number) is greater.	**Example:** $\frac{3}{7} > \frac{2}{7}$

✎*Evaluate and compare. Write < or > or =.*

1) $\frac{1}{4} + \frac{2}{4} \underline{\hspace{0.8cm}} \frac{1}{4}$

2) $\frac{3}{5} + \frac{2}{5} \underline{\hspace{0.8cm}} \frac{4}{5}$

3) $\frac{5}{7} - \frac{3}{7} \underline{\hspace{0.8cm}} \frac{6}{7}$

4) $\frac{9}{10} + \frac{7}{10} \underline{\hspace{0.8cm}} \frac{5}{10}$

5) $\frac{5}{9} - \frac{3}{9} \underline{\hspace{0.8cm}} \frac{7}{9}$

6) $\frac{10}{12} - \frac{5}{12} \underline{\hspace{0.8cm}} \frac{3}{12}$

7) $\frac{3}{8} + \frac{1}{8} \underline{\hspace{0.8cm}} \frac{1}{8}$

8) $\frac{10}{15} + \frac{4}{15} \underline{\hspace{0.8cm}} \frac{9}{15}$

9) $\frac{15}{18} - \frac{3}{18} \underline{\hspace{0.8cm}} \frac{17}{18}$

10) $\frac{17}{21} + \frac{4}{21} \underline{\hspace{0.8cm}} \frac{18}{21}$

11) $\frac{14}{16} - \frac{4}{16} \underline{\hspace{0.8cm}} \frac{12}{16}$

12) $\frac{27}{32} - \frac{11}{32} \underline{\hspace{0.8cm}} \frac{20}{32}$

13) $\frac{25}{30} + \frac{5}{30} \underline{\hspace{0.8cm}} \frac{15}{30}$

14) $\frac{25}{27} - \frac{3}{27} \underline{\hspace{0.8cm}} \frac{9}{27}$

15) $\frac{42}{45} - \frac{15}{45} \underline{\hspace{0.8cm}} \frac{30}{45}$

16) $\frac{32}{36} + \frac{15}{36} \underline{\hspace{0.8cm}} \frac{18}{36}$

Add 3 or More Fractions with Like Denominators

Helpful	- For fractions with same denominators, you only need to add the numerator (the tope number).	**Example:**
Hints	- The denominator (the bottom number) of the answer will be the same.	$\frac{3}{11} + \frac{2}{11} + \frac{5}{11} = \frac{3+2+5}{11}$ $= \frac{10}{11}$

✎ **Add fractions.**

1) $\frac{4}{7} + \frac{2}{7} + \frac{1}{7}$

8) $\frac{5}{18} + \frac{5}{18} + \frac{3}{18}$

2) $\frac{1}{5} + \frac{3}{5} + \frac{1}{5}$

9) $\frac{5}{21} + \frac{11}{21} + \frac{3}{21}$

3) $\frac{3}{9} + \frac{3}{9} + \frac{1}{9}$

10) $\frac{2}{16} + \frac{5}{16} + \frac{8}{16}$

4) $\frac{1}{4} + \frac{1}{4} + \frac{1}{4}$

11) $\frac{4}{25} + \frac{4}{25} + \frac{4}{25}$

5) $\frac{7}{15} + \frac{3}{15} + \frac{4}{15}$

12) $\frac{12}{30} + \frac{7}{30} + \frac{5}{30}$

6) $\frac{3}{12} + \frac{2}{12} + \frac{3}{12}$

13) $\frac{9}{27} + \frac{6}{27} + \frac{6}{27}$

7) $\frac{4}{10} + \frac{2}{10} + \frac{1}{10}$

14) $\frac{3}{42} + \frac{5}{42} + \frac{6}{42}$

Simplifying Fractions

Helpful		Example:
	– Evenly divide both the top and bottom of the fraction by 2, 3, 5, 7, … etc.	
Hints	– Continue until you can't go any further.	$\frac{4}{12} = \frac{2}{6} = \frac{1}{3}$

✎ *Simplify the fractions.*

1) $\frac{22}{36}$

2) $\frac{8}{10}$

3) $\frac{12}{18}$

4) $\frac{6}{8}$

5) $\frac{13}{39}$

6) $\frac{5}{20}$

7) $\frac{16}{36}$

8) $\frac{18}{36}$

9) $\frac{20}{50}$

10) $\frac{6}{54}$

11) $\frac{45}{81}$

12) $\frac{21}{28}$

13) $\frac{35}{56}$

14) $\frac{52}{64}$

15) $\frac{13}{65}$

16) $\frac{44}{77}$

17) $\frac{21}{42}$

18) $\frac{15}{36}$

19) $\frac{9}{24}$

20) $\frac{20}{80}$

21) $\frac{25}{45}$

Add Fractions with Unlike Denominators

Helpful	Adding fraction with the different denominator:	**Example:**
Hints	$\dfrac{a}{b} + \dfrac{c}{d} = \dfrac{ad + cb}{bd}$	$\dfrac{3}{4} + \dfrac{2}{3} = \dfrac{9 + 8}{12} = \dfrac{17}{12}$

✎**Add fractions.**

1) $\dfrac{2}{3} + \dfrac{1}{2}$

2) $\dfrac{3}{5} + \dfrac{1}{6}$

3) $\dfrac{5}{6} + \dfrac{1}{2}$

4) $\dfrac{3}{4} + \dfrac{5}{9}$

5) $\dfrac{2}{5} + \dfrac{1}{6}$

6) $\dfrac{3}{7} + \dfrac{1}{3}$

7) $\dfrac{3}{4} + \dfrac{2}{5}$

8) $\dfrac{2}{3} + \dfrac{1}{5}$

9) $\dfrac{16}{25} + \dfrac{3}{20}$

10) $\dfrac{2}{7} + \dfrac{1}{2}$

11) $\dfrac{3}{11} + \dfrac{2}{5}$

12) $\dfrac{1}{3} + \dfrac{1}{15}$

Subtract Fractions with Unlike Denominators

Helpful	Subtracting fractions with the different denominators:	**Example:**
Hints	$$\frac{a}{b} - \frac{c}{d} = \frac{ad - cb}{bd}$$	$$\frac{5}{12} - \frac{2}{6} = \frac{5 - 4}{12} = \frac{1}{12}$$

✎ *Subtract fractions.*

1) $\dfrac{4}{5} - \dfrac{1}{3}$

5) $\dfrac{3}{7} - \dfrac{3}{14}$

9) $\dfrac{1}{2} - \dfrac{1}{9}$

2) $\dfrac{3}{5} - \dfrac{3}{7}$

6) $\dfrac{4}{15} - \dfrac{1}{10}$

10) $\dfrac{13}{25} - \dfrac{1}{5}$

3) $\dfrac{1}{2} - \dfrac{1}{3}$

7) $\dfrac{13}{18} - \dfrac{2}{3}$

11) $\dfrac{1}{3} - \dfrac{1}{27}$

4) $\dfrac{8}{9} - \dfrac{3}{5}$

8) $\dfrac{5}{8} - \dfrac{2}{5}$

12) $\dfrac{11}{20} - \dfrac{2}{7}$

Add Fractions with Denominators of 10 and 100

Helpful	Adding fractions with the different denominators:	**Example:**
Hints	$$\frac{a}{b} + \frac{c}{d} = \frac{ad + cb}{bd}$$	$$\frac{7}{10} + \frac{20}{100} = \frac{70 + 20}{100}$$ $$= \frac{90}{100}$$

✎ *Add fractions.*

1) $\dfrac{5}{10} + \dfrac{20}{100}$

2) $\dfrac{2}{10} + \dfrac{35}{100}$

3) $\dfrac{25}{100} + \dfrac{6}{10}$

4) $\dfrac{73}{100} + \dfrac{1}{10}$

5) $\dfrac{68}{100} + \dfrac{2}{10}$

6) $\dfrac{4}{10} + \dfrac{40}{100}$

7) $\dfrac{80}{100} + \dfrac{1}{10}$

8) $\dfrac{50}{100} + \dfrac{3}{10}$

9) $\dfrac{59}{100} + \dfrac{3}{10}$

10) $\dfrac{7}{10} + \dfrac{12}{100}$

11) $\dfrac{9}{10} + \dfrac{10}{100}$

12) $\dfrac{40}{100} + \dfrac{3}{10}$

13) $\dfrac{36}{100} + \dfrac{4}{10}$

14) $\dfrac{27}{100} + \dfrac{6}{10}$

15) $\dfrac{55}{100} + \dfrac{3}{10}$

16) $\dfrac{1}{10} + \dfrac{85}{100}$

17) $\dfrac{17}{100} + \dfrac{6}{10}$

18) $\dfrac{26}{100} + \dfrac{7}{10}$

Add and Subtract Fractions with Denominators of 10, 100, and 1000

Helpful *Hints*	Adding and subtracting fractions with the different denominators: $$\frac{a}{b} + \frac{c}{d} = \frac{ad + cb}{bd}$$ $$\frac{a}{b} - \frac{c}{d} = \frac{ad - cb}{bd}$$	**Example:** $$\frac{5}{100} - \frac{30}{1000} = \frac{50 - 30}{1000}$$ $$= \frac{20}{1000}$$

✎ Evaluate fractions.

1) $\dfrac{8}{10} - \dfrac{30}{100}$

2) $\dfrac{6}{10} + \dfrac{27}{100}$

3) $\dfrac{25}{100} + \dfrac{450}{1000}$

4) $\dfrac{73}{100} - \dfrac{320}{1000}$

5) $\dfrac{25}{100} + \dfrac{670}{1000}$

6) $\dfrac{4}{10} + \dfrac{780}{1000}$

7) $\dfrac{80}{100} - \dfrac{560}{1000}$

8) $\dfrac{78}{100} - \dfrac{6}{10}$

9) $\dfrac{820}{1000} + \dfrac{5}{10}$

10) $\dfrac{67}{100} + \dfrac{240}{1000}$

11) $\dfrac{7}{10} - \dfrac{12}{100}$

12) $\dfrac{75}{100} - \dfrac{5}{10}$

13) $\dfrac{70}{100} - \dfrac{3}{10}$

14) $\dfrac{850}{1000} - \dfrac{5}{100}$

15) $\dfrac{300}{1000} + \dfrac{12}{100}$

16) $\dfrac{780}{1000} - \dfrac{6}{10}$

17) $\dfrac{80}{100} - \dfrac{6}{10}$

18) $\dfrac{50}{100} - \dfrac{210}{1000}$

Answers of Worksheets – Chapter 14

Fraction

1) $\frac{3}{8}$

2) $\frac{4}{10}$

3) $\frac{6}{20}$

4) $\frac{4}{42}$

5) $\frac{3}{8}$

Add Fractions with Like Denominators

1) 1

2) 1

3) $\frac{9}{8}$

4) $\frac{6}{4}$

5) $\frac{7}{10}$

6) $\frac{5}{7}$

7) $\frac{8}{5}$

8) $\frac{12}{14}$

9) $\frac{16}{18}$

10) $\frac{8}{12}$

11) $\frac{10}{13}$

12) $\frac{20}{25}$

13) 1

14) $\frac{9}{20}$

15) $\frac{12}{17}$

16) $\frac{33}{32}$

17) $\frac{22}{28}$

18) $\frac{12}{20}$

19) $\frac{35}{45}$

20) $\frac{26}{36}$

21) $\frac{31}{30}$

Subtract Fractions with Like Denominators

1) $\frac{2}{5}$

2) $\frac{1}{3}$

3) $\frac{3}{9}$

4) $\frac{2}{6}$

5) $\frac{1}{10}$

6) $\frac{2}{7}$

7) $\frac{2}{8}$

8) $\frac{2}{13}$

9) $\frac{3}{10}$

10) $\frac{1}{12}$

11) $\frac{6}{21}$

12) $\frac{6}{19}$

13) $\dfrac{3}{25}$

14) $\dfrac{1}{4}$

15) $\dfrac{13}{27}$

16) $\dfrac{12}{30}$

17) $\dfrac{5}{33}$

18) $\dfrac{10}{28}$

19) $\dfrac{20}{40}$

20) $\dfrac{2}{7}$

21) $\dfrac{10}{36}$

Add and Subtract Fractions with Like Denominators

1) 1

2) $\dfrac{5}{6}$

3) $\dfrac{7}{8}$

4) $\dfrac{8}{9}$

5) $\dfrac{5}{10}$

6) $\dfrac{5}{7}$

7) 1

8) $\dfrac{2}{12}$

9) $\dfrac{21}{25}$

10) $\dfrac{2}{5}$

11) $\dfrac{2}{7}$

12) $\dfrac{1}{4}$

13) $\dfrac{5}{9}$

14) $\dfrac{3}{14}$

15) $\dfrac{3}{15}$

16) $\dfrac{2}{16}$

17) $\dfrac{5}{50}$

18) $\dfrac{3}{21}$

Compare Sums and Differences of Fractions with Like Denominators

1) $\dfrac{3}{4} > \dfrac{1}{4}$

2) $1 > \dfrac{4}{5}$

3) $\dfrac{2}{7} < \dfrac{6}{7}$

4) $\dfrac{16}{10} > \dfrac{5}{10}$

5) $\dfrac{2}{9} < \dfrac{7}{9}$

6) $\dfrac{5}{12} > \dfrac{3}{12}$

7) $\dfrac{4}{8} > \dfrac{1}{8}$

8) $\dfrac{14}{15} > \dfrac{9}{15}$

9) $\dfrac{12}{18} < \dfrac{17}{18}$

10) $1 > \dfrac{18}{21}$

11) $\dfrac{10}{16} < \dfrac{12}{16}$

12) $\dfrac{16}{32} < \dfrac{20}{32}$

13) $1 > \dfrac{15}{30}$

14) $\dfrac{22}{27} > \dfrac{9}{27}$

15) $\dfrac{27}{45} < \dfrac{30}{45}$

16) $\frac{47}{36} > \frac{18}{36}$

Add 3 or More Fractions with Like Denominators

1) 1

2) 1

3) $\frac{7}{9}$

4) $\frac{3}{4}$

5) $\frac{14}{15}$

6) $\frac{8}{12}$

7) $\frac{7}{10}$

8) $\frac{13}{18}$

9) $\frac{19}{21}$

10) $\frac{15}{16}$

11) $\frac{12}{25}$

12) $\frac{24}{30}$

13) $\frac{21}{27}$

14) $\frac{14}{42}$

Simplifying Fractions

1) $\frac{11}{18}$

2) $\frac{4}{5}$

3) $\frac{2}{3}$

4) $\frac{3}{4}$

5) $\frac{1}{3}$

6) $\frac{1}{4}$

7) $\frac{4}{9}$

8) $\frac{1}{2}$

9) $\frac{2}{5}$

10) $\frac{1}{9}$

11) $\frac{5}{9}$

12) $\frac{3}{4}$

13) $\frac{5}{8}$

14) $\frac{13}{16}$

15) $\frac{1}{5}$

16) $\frac{4}{7}$

17) $\frac{1}{2}$

18) $\frac{5}{12}$

19) $\dfrac{3}{8}$ 20) $\dfrac{1}{4}$ 21) $\dfrac{5}{9}$

Add fractions with unlike denominators

1) $\dfrac{7}{6}$ 5) $\dfrac{17}{30}$ 9) $\dfrac{79}{100}$

2) $\dfrac{23}{30}$ 6) $\dfrac{16}{21}$ 10) $\dfrac{11}{14}$

3) $\dfrac{4}{3}$ 7) $\dfrac{23}{20}$ 11) $\dfrac{37}{55}$

4) $\dfrac{47}{36}$ 8) $\dfrac{13}{15}$ 12) $\dfrac{2}{5}$

Subtract fractions with unlike denominators

1) $\dfrac{7}{15}$ 5) $\dfrac{3}{14}$ 9) $\dfrac{7}{18}$

2) $\dfrac{6}{35}$ 6) $\dfrac{1}{6}$ 10) $\dfrac{8}{25}$

3) $\dfrac{1}{6}$ 7) $\dfrac{1}{18}$ 11) $\dfrac{8}{27}$

4) $\dfrac{13}{45}$ 8) $\dfrac{9}{40}$ 12) $\dfrac{37}{140}$

Add fractions with denominators of 10 and 100

1) $\dfrac{7}{10}$ 4) $\dfrac{83}{100}$ 7) $\dfrac{9}{10}$

2) $\dfrac{11}{20}$ 5) $\dfrac{22}{25}$ 8) $\dfrac{4}{5}$

3) $\dfrac{17}{20}$ 6) $\dfrac{4}{5}$ 9) $\dfrac{89}{100}$

10) $\frac{41}{50}$

13) $\frac{19}{25}$

16) $\frac{19}{20}$

11) 1

14) $\frac{87}{100}$

17) $\frac{77}{100}$

12) $\frac{7}{10}$

15) $\frac{17}{20}$

18) $\frac{24}{25}$

Add and subtract fractions with denominators of 10, 100, and 1000

1) $\frac{50}{100}$

7) $\frac{6}{25}$

13) $\frac{2}{5}$

2) $\frac{87}{100}$

8) $\frac{9}{50}$

14) $\frac{4}{5}$

3) $\frac{7}{10}$

9) $\frac{33}{25}$

15) $\frac{21}{50}$

4) $\frac{41}{100}$

10) $\frac{91}{100}$

16) $\frac{9}{50}$

5) $\frac{23}{25}$

11) $\frac{29}{50}$

17) $\frac{1}{5}$

6) $\frac{59}{50}$

12) $\frac{1}{4}$

18) $\frac{29}{100}$

Chapter 15: Mixed Numbers

Topics that you'll learn in this chapter:

- ✓ Fractions to Mixed Numbers
- ✓ Mixed Numbers to Fractions
- ✓ Add and Subtract Mixed Numbers

Fractions to Mixed Numbers

Helpful	- Divide the numerator by the denominator.	**Example:**
Helpful	- Write down the whole number of the answer.	
Hints	- Then write down any remainder above the denominator.	$\frac{7}{6} = 1\frac{7}{6}$

✎ *Convert fractions to mixed numbers.*

1) $\dfrac{9}{4}$

2) $\dfrac{37}{5}$

3) $\dfrac{21}{6}$

4) $\dfrac{41}{10}$

5) $\dfrac{11}{2}$

6) $\dfrac{56}{10}$

7) $\dfrac{20}{12}$

8) $\dfrac{9}{5}$

9) $\dfrac{19}{5}$

10) $\dfrac{27}{10}$

11) $\dfrac{10}{6}$

12) $\dfrac{17}{8}$

13) $\dfrac{7}{2}$

14) $\dfrac{39}{4}$

15) $\dfrac{72}{10}$

16) $\dfrac{13}{3}$

17) $\dfrac{45}{8}$

18) $\dfrac{27}{5}$

Mixed Numbers to Fractions

Helpful	- Multiply the whole number part by the fraction's denominator.	**Example:**
Hints	- Add the result to the numerator. - Write that result on top of the denominator.	$2\frac{3}{4} = \frac{11}{6}$

✐**Convert to fraction.**

1) $1\frac{2}{6}$

2) $2\frac{2}{3}$

3) $5\frac{1}{3}$

4) $6\frac{4}{5}$

5) $2\frac{3}{4}$

6) $2\frac{5}{7}$

7) $3\frac{5}{9}$

8) $2\frac{9}{10}$

9) $7\frac{5}{6}$

10) $6\frac{11}{12}$

11) $8\frac{9}{20}$

12) $8\frac{2}{5}$

13) $5\frac{4}{5}$

14) $9\frac{1}{6}$

15) $3\frac{3}{4}$

16) $10\frac{2}{3}$

17) $12\frac{3}{4}$

18) $14\frac{6}{7}$

Add and Subtract Mixed Numbers

Helpful	- Add whole numbers.	**Example:**
	- Add numerators and write the result on top of the common denominator.	
Hints	- If the answer is an improper fraction (the denominator is smaller than numerator), reduce the fraction into a mixed number.	
	- Simplify if necessary.	

✎ Add mixed numbers.

1) $5\frac{2}{9} + 8\frac{1}{2}$

2) $4\frac{1}{2} + 6\frac{4}{5}$

3) $6\frac{1}{3} + 7\frac{1}{4}$

4) $5\frac{1}{2} + 8\frac{1}{3}$

5) $5\frac{1}{3} - 1\frac{2}{3}$

6) $7\frac{3}{20} - 1\frac{3}{5}$

7) $7\frac{5}{9} - 2\frac{7}{9}$

8) $4\frac{4}{5} - 2\frac{9}{10}$

9) $5\frac{23}{25} - 1\frac{12}{25}$

10) $6\frac{2}{7} + 4\frac{1}{2}$

11) $3\frac{3}{8} + 2\frac{1}{8}$

12) $6\frac{2}{7} + 2\frac{1}{5}$

Answers of Worksheets – Chapter 15

Fractions to Mixed Numbers

1) $2\frac{1}{4}$

2) $7\frac{2}{5}$

3) $3\frac{1}{2}$

4) $4\frac{1}{10}$

5) $5\frac{1}{2}$

6) $5\frac{3}{5}$

7) $1\frac{2}{3}$

8) $1\frac{4}{5}$

9) $3\frac{4}{5}$

10) $2\frac{7}{10}$

11) $1\frac{2}{3}$

12) $2\frac{1}{8}$

13) $3\frac{1}{2}$

14) $9\frac{3}{4}$

15) $7\frac{1}{5}$

16) $4\frac{1}{3}$

17) $5\frac{5}{8}$

18) $5\frac{2}{5}$

Mixed Numbers to Fractions

1) $\frac{4}{3}$

2) $\frac{8}{3}$

3) $\frac{16}{3}$

4) $\frac{34}{5}$

5) $\frac{11}{4}$

6) $\frac{19}{7}$

7) $\frac{32}{9}$

8) $\frac{29}{10}$

9) $\frac{47}{6}$

10) $\frac{83}{12}$

11) $\frac{169}{20}$

12) $\frac{42}{5}$

13) $\frac{29}{5}$

14) $\frac{55}{6}$

15) $\frac{15}{4}$

16) $\dfrac{32}{3}$

17) $\dfrac{51}{4}$

18) $\dfrac{104}{7}$

Add and Subtract Mixed Numbers with Like Denominators

1) $13\dfrac{13}{18}$

2) $11\dfrac{3}{10}$

3) $13\dfrac{7}{12}$

4) $13\dfrac{5}{6}$

5) $3\dfrac{2}{3}$

6) $5\dfrac{11}{20}$

7) $4\dfrac{7}{9}$

8) $1\dfrac{9}{10}$

9) $4\dfrac{11}{25}$

10) $10\dfrac{11}{14}$

11) $5\dfrac{1}{2}$

12) $8\dfrac{17}{35}$

Chapter 16: Decimal

Topics that you'll learn in this chapter:

- ✓ Decimal Place Value
- ✓ Ordering and Comparing Decimals
- ✓ Decimal Addition
- ✓ Decimal Subtraction

Decimal Place Value

Helpful	- You can round decimals to a number of decimal places. This makes calculation easier when the exact answer is not too important. - Remember, you'll need to find the place values. - Let's review some place values.

12.3456

1: tens
2: ones
3: tenths
4: hundredths
5: thousandths
6: ten thousandths

🖎 What place is the selected digit?

1) 1,12<u>2</u>.25

2) 2,321.3<u>2</u>

3) 4,258.91

4) 6,3<u>7</u>2.67

5) 7,131.<u>9</u>8

6) <u>5</u>,442.73

7) 1,841.8<u>9</u>

8) 5,995.<u>7</u>6

9) 8,<u>9</u>82.55

10) 1,24<u>9</u>.21

🖎 What is the value of the selected digit?

11) 3,122.3<u>1</u>

12) 1,3<u>1</u>8.66

13) 6,352.<u>2</u>5

14) 3,<u>7</u>39.16

15) 4,9<u>3</u>6.78

16) 7,62<u>5</u>.86

17) 9,313.4<u>5</u>

18) <u>2</u>,168.82

19) 8,<u>4</u>51.76

20) 2,153.<u>2</u>3

Ordering and Comparing Decimals

Helpful

Hints

- **Decimals:** is a fraction written in a special form. For example, instead of writing $\frac{1}{2}$ you can write 0.5.
- **For comparing:**
Equal to =

Less than <

Greater than >

Greater than or equal ≥

Less than or equal ≤

Example:

2.67 > 0.267

✎ *Use > = <.*

1) 0.23 __ 0. 34

2) 0.31 __ 0.37

3) 0.55 __ 0.47

4) 0.57 __ 0.59

5) 0.56 __ 0. 67

6) 0.7 __ 0.67

7) 0.96 __ 8.55

8) 0.59 __ 0.88

9) 0.5 __ 0.25

10) 0.6 __ 0.3

11) 0.75 __ 0.6

12) 0.8 __ 0.80

13) 0.59 __ 0.6

14) 0.57 __ 0.75

15) 0.9 __ 0.11

16) 0.40 __ 0.4

✎ *Order each set of integers from least to greatest.*

17) 0.4, 0.54, 0.23, 0.87, 0.36   ___, ___, ___, ___, ___, ___

18) 1.2, 2.4, 1.97, 3.65, 1.80   ___, ___, ___, ___, ___, ___

19) 2.3, 1.2, 1.9, 0.67, 0.34   ___, ___, ___, ___, ___, ___

20) 1.7, 1.2, 3.2, 4.2, 1.34, 3.55   ___, ___, ___, ___, ___, ___

Decimal Addition

Helpful	1– Line up the numbers.	Example:
	2– Add zeros to have same number of digits for both numbers.	12.22
Hints	3– Add using column addition or subtraction.	$+ \ 10.34$
		22.56

✎ *Add.*

1) $8.12 + 5.24 =$

2) $1.5 + 1.3 =$

3) $7.2 + 1.34 =$

4) $3.4 + 1.75 =$

5) $2.55 + 5.25 =$

6) $5.78 + 4.30 =$

7) $12.45 + 14.25 =$

8) $13.67 + 11.31 =$

9) $16.25 + 12.34 =$

10) $10.25 + 12.55 =$

11) $21.25 + 20.90 =$

12) $16.25 + 12.88 =$

13) $18.44 + 12.65 =$

14) $32.2 + 20.45 =$

15) $15.76 + 15.98 =$

16) $25.5 + 23.9 =$

17) $30.95 + 21.40 =$

18) $23.6 + 21.6 =$

Decimal Subtraction

Helpful *Hints*	1– Line up the numbers. 2– Add zeros to have same number of digits for both numbers. 3–Subtract using column addition or subtraction.	**Example:** 16.18 − 13.45 2.73

✎**Subtract.**

1) 6.2 − 3.54 =

2) 5.77 − 4.32 =

3) 8.66 − 6.55 =

4) 7.34 − 3.22 =

5) 4.5 − 2.1 =

6) 3.78 − 2.55 =

7) 5.98 − 4.44 =

8) 4.23 − 3.9 =

9) 16.5 − 13.12 =

10) 18.67 − 11.35 =

11) 12.98 − 10.45 =

12) 14.2 − 12.4 =

13) 20.14 − 18.2 =

14) 25.6 − 24.2 =

15) 21.88 − 20.12 =

16) 27.55 − 23.4 =

17) 31.34 − 27.21 =

18) 23.34 − 21.5 =

Answers of Worksheets – Chapter 16

Decimal Place Value

1) one
2) hundredths
3) hundredths
4) tenths
5) tenths
6) thousands
7) hundredths

8) tenths
9) hundredths
10) ones
11) 0.01
12) 10
13) 0.2
14) 700

15) 30
16) 5
17) 0.05
18) 2,000
19) 400
20) 0.2

Order and Comparing Decimals

1) <
2) <
3) >
4) <
5) <
6) >

7) <
8) <
9) >
10) >
11) >
12) =

13) <
14) <
15) >
16) =

17) 0.23, 0.36, 0.4, 0.54, 0.87
18) 1.2, 1.80, 1.97, 2.4, 3.65
19) 0.34, 0.67, 1.2, 1.9, 2.3
20) 1.2, 1.34, 1.7, 3.2, 3.55, 4.2

Decimal Addition

1) 13.36
2) 2.8
3) 8.54
4) 5.15
5) 7.80
6) 10.08

7) 26.7
8) 24.98
9) 28.59
10) 22.80
11) 42.15
12) 29.13

13) 31.09
14) 52.65
15) 31.74
16) 49.4
17) 52.35
18) 45.2

Decimal Subtraction

1) 2.66

2) 1.45

3) 2.11

4) 4.12

5) 2.4

6) 1.23

7) 1.54

8) 0.33

9) 3.38

10) 7.32

11) 2.53

12) 1.8

13) 1.94

14) 1.4

15) 1.76

16) 4.15

17) 4.13

18) 1.84

ISEE Lower Level Quantitative Practice Tests

The Independent School Entrance Exam (ISEE) is an admission test developed by the Educational Records Bureau for its member schools as part of their admission process.

ISEE Lower Level tests use a multiple-choice format and contain two Mathematics sections:

Quantitative Reasoning

There are 38 questions in the Quantitative Reasoning section and students have 35 minutes to answer the questions. This section contains word problems requiring either no calculation or simple calculation.

Mathematics Achievement

There are 30 questions in the Mathematics Achievement section and students have 30 minutes to answer the questions. Mathematics Achievement measures students' knowledge of Mathematics requiring one or more steps in calculating the answer.

In this section, there are two complete ISEE Lower Level Quantitative Reasoning and Mathematics Achievement Tests. Let your student take these tests to see what score they'll be able to receive on a real ISEE test.

Good luck!

Time to Test

Time to refine your skill with a practice examination

Take a practice ISEE Lower Level Math Test to simulate the test day experience. After you've finished, score your test using the answer key.

Before You Start

- You'll need a pencil and scratch papers to take the test.

- For each question, there are four possible answers. Choose which one is best.

- It's okay to guess. You won't lose any points if you're wrong.

- Use the answer sheet provided to record your answers.

- After you've finished the test, review the answer key to see where you went wrong.

- **Calculators are NOT allowed for the ISEE Lower Level Test.**

Good Luck

ISEE Lower Level Practice Test Answer Sheets

Remove (or photocopy) these answer sheets and use them to complete the practice tests.

ISEE Lower Level Practice Test 1

Quantitative Reasoning

1 Ⓐ Ⓑ Ⓒ Ⓓ	21 Ⓐ Ⓑ Ⓒ Ⓓ	1 Ⓐ Ⓑ Ⓒ Ⓓ	21 Ⓐ Ⓑ Ⓒ Ⓓ
2 Ⓐ Ⓑ Ⓒ Ⓓ	22 Ⓐ Ⓑ Ⓒ Ⓓ	2 Ⓐ Ⓑ Ⓒ Ⓓ	22 Ⓐ Ⓑ Ⓒ Ⓓ
3 Ⓐ Ⓑ Ⓒ Ⓓ	23 Ⓐ Ⓑ Ⓒ Ⓓ	3 Ⓐ Ⓑ Ⓒ Ⓓ	23 Ⓐ Ⓑ Ⓒ Ⓓ
4 Ⓐ Ⓑ Ⓒ Ⓓ	24 Ⓐ Ⓑ Ⓒ Ⓓ	4 Ⓐ Ⓑ Ⓒ Ⓓ	24 Ⓐ Ⓑ Ⓒ Ⓓ
5 Ⓐ Ⓑ Ⓒ Ⓓ	25 Ⓐ Ⓑ Ⓒ Ⓓ	5 Ⓐ Ⓑ Ⓒ Ⓓ	25 Ⓐ Ⓑ Ⓒ Ⓓ
6 Ⓐ Ⓑ Ⓒ Ⓓ	26 Ⓐ Ⓑ Ⓒ Ⓓ	6 Ⓐ Ⓑ Ⓒ Ⓓ	26 Ⓐ Ⓑ Ⓒ Ⓓ
7 Ⓐ Ⓑ Ⓒ Ⓓ	27 Ⓐ Ⓑ Ⓒ Ⓓ	7 Ⓐ Ⓑ Ⓒ Ⓓ	27 Ⓐ Ⓑ Ⓒ Ⓓ
8 Ⓐ Ⓑ Ⓒ Ⓓ	28 Ⓐ Ⓑ Ⓒ Ⓓ	8 Ⓐ Ⓑ Ⓒ Ⓓ	28 Ⓐ Ⓑ Ⓒ Ⓓ
9 Ⓐ Ⓑ Ⓒ Ⓓ	29 Ⓐ Ⓑ Ⓒ Ⓓ	9 Ⓐ Ⓑ Ⓒ Ⓓ	29 Ⓐ Ⓑ Ⓒ Ⓓ
10 Ⓐ Ⓑ Ⓒ Ⓓ	30 Ⓐ Ⓑ Ⓒ Ⓓ	10 Ⓐ Ⓑ Ⓒ Ⓓ	30 Ⓐ Ⓑ Ⓒ Ⓓ
11 Ⓐ Ⓑ Ⓒ Ⓓ	31 Ⓐ Ⓑ Ⓒ Ⓓ	11 Ⓐ Ⓑ Ⓒ Ⓓ	
12 Ⓐ Ⓑ Ⓒ Ⓓ	32 Ⓐ Ⓑ Ⓒ Ⓓ	12 Ⓐ Ⓑ Ⓒ Ⓓ	
13 Ⓐ Ⓑ Ⓒ Ⓓ	33 Ⓐ Ⓑ Ⓒ Ⓓ	13 Ⓐ Ⓑ Ⓒ Ⓓ	
14 Ⓐ Ⓑ Ⓒ Ⓓ	34 Ⓐ Ⓑ Ⓒ Ⓓ	14 Ⓐ Ⓑ Ⓒ Ⓓ	
15 Ⓐ Ⓑ Ⓒ Ⓓ	35 Ⓐ Ⓑ Ⓒ Ⓓ	15 Ⓐ Ⓑ Ⓒ Ⓓ	
16 Ⓐ Ⓑ Ⓒ Ⓓ	36 Ⓐ Ⓑ Ⓒ Ⓓ	16 Ⓐ Ⓑ Ⓒ Ⓓ	
17 Ⓐ Ⓑ Ⓒ Ⓓ	37 Ⓐ Ⓑ Ⓒ Ⓓ	17 Ⓐ Ⓑ Ⓒ Ⓓ	
18 Ⓐ Ⓑ Ⓒ Ⓓ	38 Ⓐ Ⓑ Ⓒ Ⓓ	18 Ⓐ Ⓑ Ⓒ Ⓓ	
19 Ⓐ Ⓑ Ⓒ Ⓓ		19 Ⓐ Ⓑ Ⓒ Ⓓ	
20 Ⓐ Ⓑ Ⓒ Ⓓ		20 Ⓐ Ⓑ Ⓒ Ⓓ	

Mathematics Achievement

ISEE Lower Level Practice Test 2

Quantitative Reasoning

1 Ⓐ Ⓑ Ⓒ Ⓓ
2 Ⓐ Ⓑ Ⓒ Ⓓ
3 Ⓐ Ⓑ Ⓒ Ⓓ
4 Ⓐ Ⓑ Ⓒ Ⓓ
5 Ⓐ Ⓑ Ⓒ Ⓓ
6 Ⓐ Ⓑ Ⓒ Ⓓ
7 Ⓐ Ⓑ Ⓒ Ⓓ
8 Ⓐ Ⓑ Ⓒ Ⓓ
9 Ⓐ Ⓑ Ⓒ Ⓓ
10 Ⓐ Ⓑ Ⓒ Ⓓ
11 Ⓐ Ⓑ Ⓒ Ⓓ
12 Ⓐ Ⓑ Ⓒ Ⓓ
13 Ⓐ Ⓑ Ⓒ Ⓓ
14 Ⓐ Ⓑ Ⓒ Ⓓ
15 Ⓐ Ⓑ Ⓒ Ⓓ
16 Ⓐ Ⓑ Ⓒ Ⓓ
17 Ⓐ Ⓑ Ⓒ Ⓓ
18 Ⓐ Ⓑ Ⓒ Ⓓ
19 Ⓐ Ⓑ Ⓒ Ⓓ
20 Ⓐ Ⓑ Ⓒ Ⓓ

21 Ⓐ Ⓑ Ⓒ Ⓓ
22 Ⓐ Ⓑ Ⓒ Ⓓ
23 Ⓐ Ⓑ Ⓒ Ⓓ
24 Ⓐ Ⓑ Ⓒ Ⓓ
25 Ⓐ Ⓑ Ⓒ Ⓓ
26 Ⓐ Ⓑ Ⓒ Ⓓ
27 Ⓐ Ⓑ Ⓒ Ⓓ
28 Ⓐ Ⓑ Ⓒ Ⓓ
29 Ⓐ Ⓑ Ⓒ Ⓓ
30 Ⓐ Ⓑ Ⓒ Ⓓ
31 Ⓐ Ⓑ Ⓒ Ⓓ
32 Ⓐ Ⓑ Ⓒ Ⓓ
33 Ⓐ Ⓑ Ⓒ Ⓓ
34 Ⓐ Ⓑ Ⓒ Ⓓ
35 Ⓐ Ⓑ Ⓒ Ⓓ
36 Ⓐ Ⓑ Ⓒ Ⓓ
37 Ⓐ Ⓑ Ⓒ Ⓓ
38 Ⓐ Ⓑ Ⓒ Ⓓ

Mathematics Achievement

1 Ⓐ Ⓑ Ⓒ Ⓓ
2 Ⓐ Ⓑ Ⓒ Ⓓ
3 Ⓐ Ⓑ Ⓒ Ⓓ
4 Ⓐ Ⓑ Ⓒ Ⓓ
5 Ⓐ Ⓑ Ⓒ Ⓓ
6 Ⓐ Ⓑ Ⓒ Ⓓ
7 Ⓐ Ⓑ Ⓒ Ⓓ
8 Ⓐ Ⓑ Ⓒ Ⓓ
9 Ⓐ Ⓑ Ⓒ Ⓓ
10 Ⓐ Ⓑ Ⓒ Ⓓ
11 Ⓐ Ⓑ Ⓒ Ⓓ
12 Ⓐ Ⓑ Ⓒ Ⓓ
13 Ⓐ Ⓑ Ⓒ Ⓓ
14 Ⓐ Ⓑ Ⓒ Ⓓ
15 Ⓐ Ⓑ Ⓒ Ⓓ
16 Ⓐ Ⓑ Ⓒ Ⓓ
17 Ⓐ Ⓑ Ⓒ Ⓓ
18 Ⓐ Ⓑ Ⓒ Ⓓ
19 Ⓐ Ⓑ Ⓒ Ⓓ
20 Ⓐ Ⓑ Ⓒ Ⓓ

21 Ⓐ Ⓑ Ⓒ Ⓓ
22 Ⓐ Ⓑ Ⓒ Ⓓ
23 Ⓐ Ⓑ Ⓒ Ⓓ
24 Ⓐ Ⓑ Ⓒ Ⓓ
25 Ⓐ Ⓑ Ⓒ Ⓓ
26 Ⓐ Ⓑ Ⓒ Ⓓ
27 Ⓐ Ⓑ Ⓒ Ⓓ
28 Ⓐ Ⓑ Ⓒ Ⓓ
29 Ⓐ Ⓑ Ⓒ Ⓓ
30 Ⓐ Ⓑ Ⓒ Ⓓ

ISEE Lower Level
Practice Test 1

Quantitative Reasoning

38 questions

Total time for this test: 35 Minutes

You may NOT use a calculator for this test.

1) Which of the following is greater than $\frac{12}{8}$?

A. $\frac{1}{2}$

B. $\frac{5}{2}$

C. $\frac{4}{3}$

D. 1

2) If $\frac{1}{3}$ of a number is greater than 8, the number must be .....

A. Less than 4

B. Equal to 16

C. Equal to 24

D. Greater than 24

3) If $4 \times (M + N) = 20$ and M is greater than 0, then N could Not be .....

A. 1

B. 2

C. 3

D. 5

4) Which of the following is closest to 5.03?

A. 6

B. 5.5

C. 5

D. 5.4

5) At a Zoo, the ratio of lions to tigers is 10 to 6. Which of the following could NOT be the total number of lions and tigers in the zoo?

A. 64

B. 80

C. 98

D. 104

6) In the multiplication bellow, A represents which digit?
$$14 \times 3A2 = 4,788$$

A. 2

B. 3

C. 4

D. 6

7) If N is an even number, which of the following is always an odd number?

A. $\frac{N}{2}$

B. $N + 4$

C. $2N$

D. $N + 1$

8) $8.9 - 4.08$ is closest to which of the following.

A. 4.1

B. 4.8

C. 6

D. 8

$$x = 2,456 \qquad y = 259$$

9) Numbers x and y are shown above. How many times larger is the value of digit 5 in the number x, than the value of digit 5 in the number y?

A. 1

B. 10

C. 100

D. 1,000

10) If 5 added to a number, the sum is 20. If the same number added to 25, the answer is

A. 30

B. 35

C. 40

D. 45

11) $\dfrac{2+5+6\times1+1}{3+5} = ?$

A. $\dfrac{15}{8}$

B. $\dfrac{4}{8}$

C. $\dfrac{7}{4}$

D. $\dfrac{6}{8}$

12) $7 \times 4 \times 12 \times 3$ is equal to the product of 28 and

A. 3

B. 12

C. 24

D. 36

13) If 20 can be divided by both 4 and x without leaving a remainder, then 20 can also be divided by which of the following?

A. $x + 4$

B. $2x - 4$

C. $x - 2$

D. $x \times 4$

14) Use the equations below to answer the question:

$$x - 12 = 18$$
$$16 + y = 21$$

What is the value of $x + y$?

A. 9

B. 10

C. 11

D. 12

15) Which of the following expressions has the same value as $\frac{5}{4} \times \frac{6}{2}$?

A. $\frac{6 \times 3}{4}$

B. $\frac{6 \times 2}{4}$

C. $\frac{5 \times 6}{4}$

D. $\frac{5 \times 3}{4}$

16) When 5 is added to three times number N, the result is 41. Then N is ....

A. 11

B. 12

C. 14

D. 16

E. 18

17) At noon, the temperature was 15 degrees. By midnight, it had dropped another 20 degrees. What was the temperature at midnight?

A. 10 degrees above zero

B. 10 degrees below zero

C. 5 degrees above zero

D. 5 degrees below zero

18) If a triangle has a base of 5 cm and a height of 8 cm, what is the area of the triangle?

A. $15\ cm^2$

B. $20\ cm^2$

C. $40\ cm^2$

D. $45\ cm^2$

19) Which formula would you use to find the area of a square?

A. $length \times width \times height$

B. $\frac{1}{2} base \times height$

C. $length \times width$

D. $side \times side$

20) What is the next number in this sequence?   2, 5, 9, 14, 20, …

A. 27

B. 26

C. 25

D. 21

21) What is the average of the following numbers?   6, 10, 12, 23, 45

A. 19

B. 19.2

C. 19.5

D. 20

22) If there are 8 red balls and 12 blue balls in a basket, what is the probability that John will pick out a red ball from the basket?

A. $\frac{18}{10}$

B. $\frac{2}{5}$

C. $\frac{2}{10}$

D. $\frac{3}{5}$

23) How many lines of symmetry does an equilateral triangle have?

A. 5

B. 4

C. 3

D. 2

24) What is %10 of 200?

A. 10

B. 20

C. 30

D. 40

25) Which of the following statement is False?

A. $2 \times 2 = 4$

B. $(4 + 1) \times 5 = 25$

C. $6 \div (3 - 1) = 1$

D. $6 \times (4 - 2) = 12$

26) If all the sides in the following figure are of equal length and length of one side is 4, what is the perimeter of the figure?

A. 15

B. 18

C. 20

D. 24

27) $\frac{4}{5} - \frac{3}{5} = ?$

A. 0.3

B. 0.35

C. 0.2

D. 0.25

28) If $N = 2$ and $\frac{64}{N} + 4 = \square$, then $\square$ = ...

A. 30

B. 32

C. 34

D. 36

29) Four people can paint 4 houses in 10 days. How many people are needed to paint 8 houses in 5 days?

A. 6

B. 8

C. 12

D. 16

30) What is the median of these numbers? 4, 9, 13, 8, 15, 18, 5

A. 8

B. 9

C. 13

D. 15

The result of a research shows the number of men and women in four cities of a country.

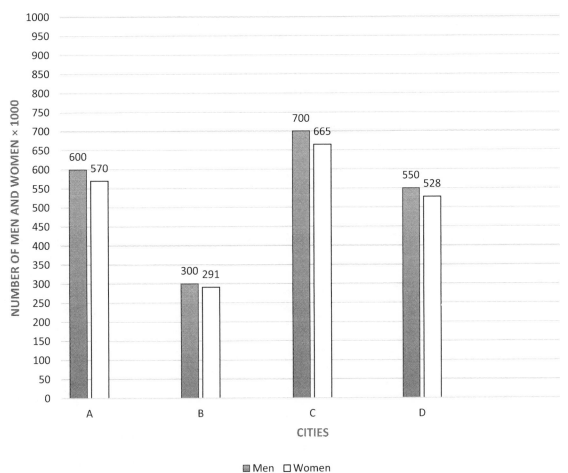

Number of men and women in four cities

31) What is the difference of the population of men in the biggest city and in the smallest city?

A. 200

B. 300

C. 400

D. 500

32) What is 5,231.48245 rounded to the nearest tenth?

A. 5,231.482

B. 5,231.5

C. 5,231

D. 5,231.48

33) $18a + 22 = 40, a = ?$

A. 1

B. 4

C. 11

D. 12

34) Two angles of a triangle measure 35 and 65. What is the measure of third angle?

A. 50

B. 60

C. 70

D. 80

35) A woman weighs 135 pounds. She gains 15 pounds one month and 8 pounds the next month. What is her new weight?

A. 152 Pounds

B. 146 Pounds

C. 158 Pounds

D. 138 Pounds

36) In a basket, there are equal numbers of red, white, yellow, and purple cards. Which of the following could be the number of cards in the basket?

A. 123

B. 82

C. 66

D. 56

37) Jim types 88 words per minute. How many words does he type in 15 seconds?

A. 15

B. 18

C. 22

D. 25

38) Which of the following is NOT equal to $\frac{2}{7}$?

A. $\frac{22}{77}$

B. $\frac{8}{28}$

C. $\frac{18}{63}$

D. $\frac{12}{48}$

IF YOU FINISH BEFORE TIME IS CALLED, YOU MAY CHECK YOUR WORK ON THIS SECTION ONLY. DO NOT TURN TO ANY OTHER SECTION IN THE TEST.

STOP

ISEE Lower Level

Practice Test 1

Mathematics Achievement

30 questions

Total time for this test: 30 Minutes

You may NOT use a calculator for this test.

1) $\frac{1}{5} + \frac{3}{4} =$

   A. $\frac{4}{9}$

   B. $\frac{3}{9}$

   C. $\frac{3}{4}$

   D. $\frac{19}{20}$

2) What's the least common multiple (LCM) of 8 and 14?

   A. 8 and 14 have no common multiples

   B. 112

   C. 96

   D. 56

3) Which of the following is NOT a factor of 50?

   A. 5

   B. 2

   C. 10

   D. 15

4) While at work, Emma checks her email once every 90 minutes. In 9-hour, how many times does she check her email?

A. 4 Times

B. 5 Times

C. 7 Times

D. 6 Times

5) What is 8,923.2769 rounded to the nearest tenth?

A. 8,923.3

B. 8,923.277

C. 8,923

D. 8,923.27

6) Which of the following fractions is the largest?

A. $\frac{3}{4}$

B. $\frac{2}{5}$

C. $\frac{7}{9}$

D. $\frac{2}{3}$

7) A bag contains 18 balls: two green, five black, eight blue, a brown, a red and one white. If 17 balls are removed from the bag at random, what is the probability that a brown ball has been removed?

A. $\dfrac{1}{9}$

B. $\dfrac{1}{6}$

C. $\dfrac{16}{17}$

D. $\dfrac{17}{18}$

8) From last year, the price of gasoline has increased from \$1.25 per gallon to \$1.75 per gallon. The new price is what percent of the original price?

A. 72%

B. 120%

C. 140%

D. 160%

9) Emma purchased a computer for \$530.40. The computer is regularly priced at \$624. What was the percent discount Emma received on the computer?

A. 12%

B. 15%

C. 20%

D. 25%

10) In the given diagram, the height is 9 cm. what is the area of the triangle?

A. 23 cm²

B. 46 cm²

C. 126 cm²

D. 252 cm²

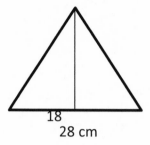

18

28 cm

11) Two angles of a triangle measure 51 and 47. What is the measure of the third angle?

A. 82

B. 99

C. 102

D. 262

12) If a rectangular swimming pool has a perimeter of 112 feet and is 22 feet wide, what is its area?

A. 1,496

B. 900

C. 840

D. 748

13) Mike is 7.5 miles ahead of Julia running at 5.5 miles per hour and Julia is running at the speed of 6 miles per hour. How long does it take Julia to catch Mike?

A. 2 hours

B. 5.5 hours

C. 7.5 hours

D. 15 hours

14) Julie gives 8 pieces of candy to each of her friends. If Julie gives all her candy away, which amount of candy could have been the amount she distributed?

A. 187

B. 216

C. 243

D. 223

15) A taxi driver earns $9 per 1-hour work. If he works 10 hours a day and in 1 hour he uses 2-liters petrol with price $1 for 1-liter. How much money does he earn in one day?

A. $90

B. $88

C. $70

D. $60

16) Convert 0.025 to a percent.

A. 0.03%

B. 0.25%

C. 2.50%

D. 25%

17) The number 0.04 can also represented by which of the following?

A. $\dfrac{4}{10}$

B. $\dfrac{4}{100}$

C. $\dfrac{4}{1,000}$

D. $\dfrac{4}{10,000}$

18) The width of a box is one third of its length. The height of the box is one third of its width.

   If the length of the box is 27 cm, what is the volume of the box?

A. 81 cm^3

B. 162 cm^3

C. 243 cm^3

D. 729 cm^3

19) 125 students took an exam and 25 of them failed. What percent of the students passed the exam?

A. 20 %

B. 40 %

C. 60 %

D. 80 %

20) $\begin{array}{r} 37 \text{ hr. } 25 \text{ min.} \\ - 23 \text{ hr. } 38 \text{ min.} \\ \hline \end{array}$

A. 12 hr. 57 min.

B. 12 hr. 47 min.

C. 13 hr. 47 min.

D. 13 hr. 57 min.

21) Which of the following is an obtuse angle?

A. 116∘

B. 80∘

C. 68∘

D. 25∘

Use the following table to answer question below.

DANIEL'S BIRD-WATCHING PROJECT	
DAY	NUMBER OF RAPTORS SEEN
Monday	?
Tuesday	9
Wednesday	14
Thursday	12
Friday	5
MEAN	10

22) This table shows the data Daniel collects while watching birds for one week. How many raptors did Daniel see on Monday?

A. 10

B. 11

C. 12

D. 13

23) In the following figure, the shaded squares are what fractional part of the whole set of squares?

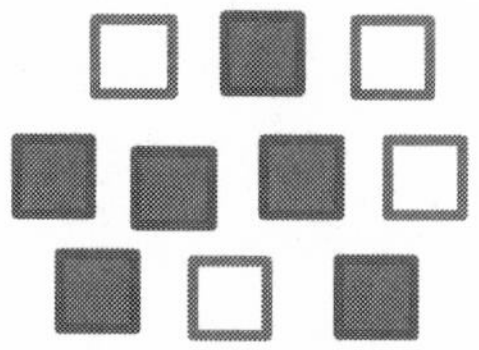

A. $\frac{1}{2}$

B. $\frac{5}{8}$

C. $\frac{2}{3}$

D. $\frac{3}{5}$

24) If a box contains red and blue balls in ratio of 2 : 3 red to blue, how many red balls are there if 90 blue balls are in the box?

A. 40

B. 60

C. 80

D. 82

25) A shirt costing \$200 is discounted 15%. After a month, the shirt is discounted another 15%. Which of the following expressions can be used to find the selling price of the shirt?

A. $(200)(0.70)$

B. $(200) - 200(0.30)$

C. $(200)(0.15) - (200)(0.15)$

D. $(200)(0.85)(0.85)$

26) Emma draws a shape on her paper. The shape has four sides. It has only one pair of parallel sides. What shape does Emma draw?

A. Parallelogram

B. Rectangle

C. Square

D. Trapezoid

27) If $A = 20$, then which of the following equations are correct?

A. $A + 20 = 40$

B. $A \div 20 = 40$

C. $20 \times A = 40$

D. $A - 20 = 40$

28) Joe makes $4.75 per hour at his work. If he works 8 hours, how much money will he earn?

A. $32.00

B. $34.75

C. $36.50

D. $38.00

29) In a classroom of 44 students, 18 are male. About what percentage of the class is female?

 A. 63%

 B. 51%

 C. 59%

 D. 53%

30) Nancy ordered 18 pizzas. Each pizza has 8 slices. How many slices of pizza did Nancy ordered?

A. 124

B. 144

C. 156

D. 180

IF YOU FINISH BEFORE TIME IS CALLED, YOU MAY CHECK YOUR WORK ON THIS SECTION ONLY. DO NOT TURN TO ANY OTHER SECTION IN THE TEST. STOP

ISEE Lower Level

Practice Test 2

Quantitative Reasoning

38 questions

Total time for this test: 35 Minutes

You may NOT use a calculator for this test.

1) $\frac{8}{2} - \frac{3}{2} = ?$

A. 1

B. 1.5

C. 2

D. 2.5

2) If $48 = 3 \times N + 12$, then $N = \ldots$

A. 8

B. 12

C. 14

D. 15

3) When 3 is added to four times a number N, the result is 23. Which of the following equations represents this statement?

A. $4 + 3N = 23$

B. $23N + 4 = 3$

C. $4N + 3 = 23$

D. $4N + 23 = 3$

4) When 78 is divided by 5, the remainder is the same as when 45 is divided by

A. 2

B. 4

C. 5

D. 7

5) John has 2,400 cards and Max has 606 cards. How many more cards does John have than Max?

A. 1,794

B. 1,798

C. 1,812

D. 1,828

6) In the following right triangle, what is the value of x?

A. 15

B. 30

C. 45

D. 60

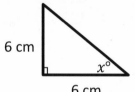

7) What is 5 percent of 480?

A. 20

B. 24

C. 30

D. 40

8) In a basket, the ratio of red marbles to blue marbles is 3 to 2. Which of the following could NOT be the total number of red and blue marbles in the basket?

A. 15

B. 32

C. 55

D. 60

9) A square has an area of $81 cm^2$. What is its perimeter?

A. $28\ cm^2$

B. $32\ cm^2$

C. $34\ cm^2$

D. $36\ cm^2$

10) Find the missing number in the sequence: 5, 8, 12, ..., 23

A. 15

B. 17

C. 18

D. 20

11) The length of a rectangle is 3 times of its width. If the length is 18, what is the perimeter of the rectangle?

A. 24

B. 30

C. 36

D. 48

12) Mary has y dollars. John has $10 more than Mary. If John gives Mary $12, then in terms of y, how much does John have now?

A. $y + 1$

B. $y + 10$

C. $y - 2$

D. $y - 1$

13) Dividing 107 by 6 leaves a remainder of

A. 1

B. 2

C. 3

D. 5

14) If $6,000 + A - 200 = 7,400$, then $A =$

A. 200

B. 600

C. 1,600

D. 2,200

15) For what price is 15 percent off the same as $75 off?

A. $200

B. $300

C. $350

D. $500

16) Which of the following fractions is less than $\frac{3}{2}$?

A. 1.4

B. $\frac{5}{2}$

C. 3

D. 2.8

17) Use the equation below to answer the question.

$$x + 3 = 6$$
$$2y = 8$$

What is the value of $y - x$?

A. 1

B. 2

C. 3

D. 4

18) If $310 - x + 116 = 225$, then $x = ?$

A. 101

B. 156

C. 201

D. 211

19) Of the following, 25 percent of $43.99 is closest to

A. $9.90

B. $10.00

C. $11.00

D. $11.50

20) Solve.

$8.08 - 5.6 =$ ….

A. 2.42

B. 2.46

C. 2.48

D. 3

21) If $500 + \square - 180 = 1{,}100$, then $\square = ?$

A. 580

B. 660

C. 700

D. 780

22) There are 60 students in a class. If the ratio of the number of girls to the total number of students in the class is $\frac{1}{6}$, which are the following is the number of boys in that class?

A. 10

B. 20

C. 25

D. 50

23) If $N \times (5 - 3) = 12$ then $N = ?$

A. 6

B. 12

C. 13

D. 14

24) If $x \blacksquare y = 3x + y - 2$, what is the value of $4 \blacksquare 12$?

A. 4

B. 18

C. 22

D. 36

25) Of the following, which number if the greatest?

A. 0.092

B. 0.8913

C. 0.8923

D. 0.8896

26) $\dfrac{7}{8} - \dfrac{3}{4} = ?$

A. 0.125

B. 0.375

C. 0.5

D. 0.625

27) Which of the following is the closest to 4.02?

A. 4

B. 4.2

C. 4.3

D. 4.5

28) Which of the following statements is False?

A. $(7 \times 2 + 14) \times 2 = 56$

B. $(2 \times 5 + 4) \div 2 = 7$

C. $3 + (3 \times 6) = 21$

D. $14 \div (2 + 5) = 5$

29) A trash container, when empty, weighs 35 pounds. If this container is filled with a load of trash that weighs 240 pounds, what is the total weight of the container and its contents?

A. 224 pounds

B. 275 pounds

C. 285 pounds

D. 325 pounds

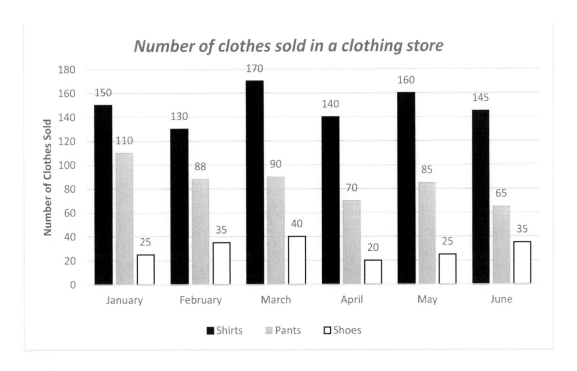

Number of clothes sold in a clothing store

30) During the six-month period shown, what is the median number of shirts per month?

A. 140

B. 147.5

C. 150

D. 170

31) A writer finishes 180 pages of his manuscript in 20 hours. How many pages is his average?

A. 18

B. 15

C. 12

D. 9

32) The distance between cities A and B is approximately 2,600 miles. If Nicole drives an average of 68 miles per hour, how many hours will it take her to drive from city A to city B?

A. approximately 41 hours

B. approximately 38 hours

C. approximately 29 hours

D. approximately 27 hours

33) $\frac{13}{25}$ is equal to:

A. 5.2

B. 0.26

C. 0.52

D. 0.13

34) $12a + 20 = 140, a = ?$

A. 12

B. 10

C. 14

D. 18

35) What is the place value of 2 in 5.7325?

A. Hundredths

B. Thousandths

C. Ten thousandths

D. Hundred thousandths

36) Which of the following is NOT a prime factor of 60?

A. 2

B. 3

C. 5

D. 6

37) Which of these numbers is equal to $\frac{25}{1,000}$?

A. 2.5

B. 0.25

C. 0.025

D. 0.0025

38) During a 24-hour day, Moe works $\frac{1}{3}$ of the time. How many hours does Moe work in that day?

A. 2

B. 4

C. 6

D. 8

STOP

ISEE Lower Level

Practice Test 2

Mathematics Achievement

30 questions

Total time for this test: 30 Minutes

You may NOT use a calculator for this test.

1) Which is sixty-five thousand, eight hundred nineteen?

   A. 65,819

   B. 650,819

   C. 605,819

   D. 658,019

2) What's the greatest common factor of the 18 and 32?

   A. 16

   B. 9

   C. 2

   D. 1

3) Which of the following is not a multiple of 5?

   A. 12

   B. 30

   C. 15

   D. 20

4) A right triangle has two short sides with lengths 6 and 8. What is the perimeter of the triangle?

A. 10

B. 18

C. 24

D. 36

5) What is the name of a rectangle with sides of equal length?

A. Hexagon

B. Octagon

C. Pentagon

D. Square

6) With what number must 5.674321 be multiplied in order to obtain the number 56,743.21?

A. 100

B. 1,000

C. 10,000

D. 100,000

7) Which expression is equal to $\frac{3}{11}$?

A. $3 - 11$

B. $3 \div 11$

C. 3×11

D. $\frac{11}{3}$

8) Lily and Ella are in a pancake–eating contest. Lily can eat two pancakes per minute, while Ella can eat $2\frac{1}{2}$ pancakes per minute. How many total pancakes can they eat in 5 minutes?

A. 9.5 Pancakes

B. 29.5 Pancakes

C. 22.5 Pancakes

D. 11.5 Pancakes

9) Which expression has a value of -8?

A. $8 - (-2) + (-18)$

B. $2 + (-3) \times (-2)$

C. $-6 \times (-6) + (-2) \times (-12)$

D. $(-2) \times (-7) + 4$

10) $0.87 + 1.4 + 3.23 = ?$

A. 2.5

B. 3.2

C. 5.5

D. 6.5

11) What is the perimeter of a rectangle that has a length of 8 inches and a width of 5 inches?

A. 13

B. 23

C. 26

D. 28

12) How many $\frac{1}{4}$ cup servings are in a package of cheese that contains $6\frac{1}{2}$ cups altogether?

A. 20

B. 22

C. 24

D. 26

13) If the following clock shows a time in the morning, what time was it 6 hours and 30 minutes ago?

A. 07:45 AM

B. 05:45 AM

C. 07:45 PM

D. 05:45 PM

14) The area of a rectangle is 72 square meters. The width is 8 meters. What is the length of the rectangle?

A. 8

B. 9

C. 10

D. 12

Use the table below to answer the question.

City Populations

City	Population
Denton	28,097
Bomberg	28,207
Windham	29,700
Sanhill	27,980

15) Which list of city populations is in order from least to greatest?

A. 28,097;   28,207;   29,700;   27,980

B. 29,700;   28,207;   28,097;   27,980

C. 27,980;   28,097;   28,207;   29,700

D. 27,980;   28,207;   28,097;   29,700

16) The temperature on Sunday at 12:00 PM was 76°F. Low temperature on the same day was 24°F cooler. Which temperature is closest to the low temperature on that day?

A. 76°F

B. 52°F

C. 51°F

D. 75°

17) $(5 + 7) \div (3^2 \div 3) =$ __

A. $\frac{5}{7}$

B. 2

C. 4

D. 12

18) Ella buys five items costing $2.26, $14.69, $2.50, $4.66, and $17.99. What is the estimated total cost of Ella's items?

A. between $25 and $30

B. between $30 and $35

C. between $35 and $40

D. between $40 and $45

19) How long is the line segment shown on the number line below?

A. 6

B. 7

C. 8

D. 9

-10 -9 -8 -7 -6 -5 -4 -3 -2 -1 0 1 2 3 4 5 6 7 8 9 10

20) There are 86 students from Riddle Elementary school at the library on Monday. The other 32 students in the school are practicing in the classroom. Which number sentence shows the total number of students in Riddle Elementary school?

A. $86 + 32$

B. $86 - 32$

C. 86×32

D. $86 \div 32$

21) $\frac{11}{19}$ is equal to:

A. 0.579

B. 5.79

C. 57.90

D. 579.00

22) Which number correctly completes the number sentence $80 \times 34 = ?$

A. 272

B. 560

C. 1,920

D. 2,720

23) What fraction of each shape is shaded?

a)

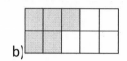
b)

A. a. $\frac{5}{16}$; b. $\frac{5}{10}$

B. a. $\frac{6}{16}$; b. $\frac{5}{16}$

C. a. $\frac{6}{16}$; b. $\frac{5}{10}$

D. a. $\frac{8}{16}$; b. $\frac{4}{10}$

24) Which statement about the number 945,382.16 is true?

A. The digit 6 has a value of (6×100)

B. The digit 4 has a value of (4×100)

C. The digit 8 has a value of (8×10)

D. The digit 5 has a value of (5×100)

25) Ella described a number using these clues:

Three – digit odd numbers that have a 6 in the hundreds place and a 3 in the tens place

Which number could fit Ella's description?

A. 627

B. 637

C. 632

D. 636

26) The following graph shows the mark of six students in mathematics. What is the mean (average) of the marks?

A. 15

B. 14.5

C. 14

D. 13.5

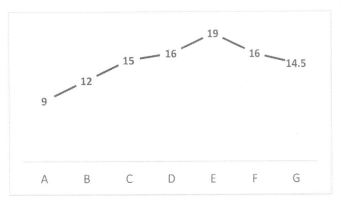

27) In the deck of cards, there are 4 spades, 3 hearts, 7 clubs, and 10 diamonds. What is the probability that William will pick out a spade?

A. $\frac{1}{6}$

B. $\frac{1}{8}$

C. $\frac{1}{9}$

D. $\frac{1}{5}$

28) Jason's favorite sports team has won 0.62 of its games this season. How can Jason express this decimal as a fraction?

A. $\frac{6}{2}$

B. $\frac{62}{10}$

C. $\frac{62}{100}$

D. $\frac{6.2}{10}$

29) Which fraction has the least value?

A. $\frac{1}{2}$

B. $\frac{3}{8}$

C. $\frac{3}{4}$

D. $\frac{9}{16}$

30) There are 365 days in a year, and 24 hours in a day. How many hours are in a year?

A. 2,190

B. 7,440

C. 7,679

D. 8,760

IF YOU FINISH BEFORE TIME IS CALLED, YOU MAY CHECK YOUR WORK ON THIS SECTION ONLY. DO NOT TURN TO ANY OTHER SECTION IN THE TEST.

STOP

ISEE Lower Level Math Practice Tests

Answers and Explanations

ISEE Lower Level Math Practice Test 1 Answer Key						
Quantitative Reasoning				**Mathematics Achievement**		

1	**B**	17	**D**	33	**A**		1	**D**	17	**B**
2	**D**	18	**B**	34	**D**		2	**D**	18	**D**
3	**D**	19	**D**	35	**C**		3	**D**	19	**D**
4	**C**	20	**A**	36	**D**		4	**D**	20	**C**
5	**C**	21	**B**	37	**C**		5	**A**	21	**A**
6	**C**	22	**B**	38	**D**		6	**C**	22	**A**
7	**D**	23	**C**				7	**D**	23	**D**
8	**B**	24	**B**				8	**C**	24	**B**
9	**A**	25	**C**				9	**B**	25	**D**
10	**C**	26	**D**				10	**C**	26	**D**
11	**C**	27	**C**				11	**A**	27	**A**
12	**D**	28	**D**				12	**D**	28	**D**
13	**D**	29	**D**				13	**D**	29	**C**
14	**C**	30	**B**				14	**B**	30	**B**
15	**D**	31	**C**				15	**C**		
16	**B**	32	**B**				16	**C**		

ISEE Lower Level Math Practice Test 2 Answer Key

Quantitative Reasoning   Mathematics Achievement

Quantitative Reasoning						Mathematics Achievement				
1	D		17	A		33	C		1	A		17	C
2	B		18	C		34	B		2	C		18	D
3	C		19	C		35	B		3	A		19	D
4	D		20	D		36	D		4	C		20	A
5	A		21	D		37	C		5	D		21	A
6	C		22	D		38	D		6	C		22	D
7	B		23	A					7	B		23	C
8	B		24	C					8	C		24	C
9	D		25	C					9	A		25	B
10	B		26	A					10	C		26	B
11	D		27	A					11	C		27	A
12	C		28	D					12	D		28	C
13	D		29	B					13	C		29	B
14	C		30	B					14	B		30	D
15	D		31	D					15	C			
16	A		32	B					16	B			

Score Your Test

ISEE scores are broken down by four sections: Verbal Reasoning, Reading Comprehension, Quantitative Reasoning, and Mathematics Achievement. A sum of the ALL sections is also reported. The Essay section is scored separately.

For the Lower Level ISEE, the score range is 760 to 940, the lowest possible score a student can earn is 760 and the highest score is 940 for each section. A student receives 1 point for every correct answer. There is no penalty for wrong or skipped questions.

The total scaled score for a Lower Level ISEE test is the sum of the scores for all sections. A student will also receive a percentile score of between 1-99% that compares that student's test scores with those of other test takers of same grade and gender from the past 3 years. When a student receives her/his score, the percentile score is also be broken down into a stanine and the stanines are ranging from 1–9. Most schools accept students with scores of 5–9. The ideal candidate has scores of 6 or higher.

Percentile Rank	Stanine
1 – 3	1
4 – 10	2
11- 22	3
23 - 39	4
40 – 59	5
60 – 76	6
77- 88	7
89 – 95	8
96 - 99	9

The following charts provide an estimate of students ISEE percentile rankings for the practice tests, compared against other students taking these tests. Keep in mind that these percentiles are estimates only, and your actual ISEE percentile will depend on the specific group of students taking the exam in your year.

ISEE Lower Level Quantitative Reasoning Percentiles			
Grade Applying to	25th Percentile	50th Percentile	75th Percentile
Grade 5	825	840	860
Grade 6	838	855	870

ISEE Lower Level Mathematics Achievement Percentiles			
Grade Applying to	25th Percentile	50th Percentile	75th Percentile
Grade 5	830	850	865
Grade 6	855	865	978

Use the next table to convert ISEE Lower level raw score to scaled score for application to grade 5 and grade 6.

ISEE Upper Level Scaled Scores									
Raw Score	Quantitative Reasoning		Mathematics Achievement		Raw Score	Quantitative Reasoning		Mathematics Achievement	
	Grade 5	Grade 6	Grade 5	Grade 6		Grade 5	Grade 6	Grade 5	Grade 6
0	760	760	760	760	26	900	885	900	890
1	770	765	770	765	27	905	890	910	900
2	780	770	780	770	28	910	895	925	920
3	790	775	790	775	29	910	900	935	930
4	800	780	800	780	30	915	905	940	940
5	810	785	810	785	31	920	910		
6	820	790	820	790	32	925	915		
7	825	795	825	795	33	930	920		
8	830	800	830	800	34	930	925		
9	835	805	835	805	35	930	925		
10	840	810	840	810	36	935	930		
11	845	815	845	815	37	935	935		
12	850	820	850	820	38	940	940		
13	855	825	855	825					
14	860	830	855	830					
15	865	835	860	835					
16	870	840	860	840					
17	875	845	865	840					
18	880	845	865	845					
19	880	850	870	845					
20	885	855	870	850					
21	885	860	875	850					
22	890	865	875	855					
23	890	870	875	855					
24	895	875	880	860					
25	895	880	890	880					

ISEE Lower Level Test 1 Practice Tests

Answers and Explanations

Quantitative Reasoning

1) Choice B is correct.

$\frac{12}{8} = \frac{3}{2} = 1.5$, the only choice that is greater than 1.5 is $\frac{5}{2}$. (Recall that $\frac{4}{3} = 1.33..$)

$$\frac{5}{2} = 2.5\, , 2.5 > 1.5$$

2) Choice D is correct.

If $\frac{1}{3}$ of a number is greater than 8, the number must be greater than 24.

$$\frac{1}{3}x > 8 \rightarrow \text{multiply both sides of the inequality by 3, then: } x > 24$$

3) Choice D is correct.

$4 \times (M + N) = 20$, then $M + N = 5$. $M > 0 \rightarrow N$ *could not be* 5

4) Choice C is correct.

The closest to 5.03 is 5 in the choices provided.

5) Choice C is correct.

The ratio of lions to tigers is 10 to 6 or 5 to 3 at the zoo. Therefore, total number of lions and tigers must be divisible by 8.

$$5 + 3 = 8$$

From the numbers provided, only 98 is not divisible by 8.

6) Choice C is correct.

A represents digit 4 in the multiplication.

$$14 \times 342 = 4,788$$

7) Choice D is correct.

N is even. Let's choose 2 and 4 for N. Now, let's review the choices provided.

A) $\frac{N}{2} = \frac{2}{2} = 1, \qquad \frac{N}{2} = \frac{4}{2} = 2,$ One result is odd and the other one is even.

B) $N + 4 = 2 + 4 = 6, 4 + 4 = 8$ Both results are even.

C) $2N = 2 \times 2 = 4, 4 \times 2 = 8$ Both results are even.

D) $N + 1 = 2 + 1 = 3, 4 + 1 = 5$ Both results are odd.

8) Choice B is correct.

$8.9 - 4.08 = 4.82$, which is closest to 4.8

9) Choice A is correct.

The value of digit 5 in both numbers x and y are in the tens place. Therefore, they have the same value.

10) Choice C is correct.

Let x be the number. Then:

$$5 + x = 20 \rightarrow x = 15 \rightarrow 15 + 25 = 40$$

11) Choice C is correct.

$$\frac{2 + 5 + 6 \times 1 + 1}{5 + 3} = \frac{14}{8} = \frac{7}{4}$$

12) Choice D is correct.

$7 \times 4 \times 12 \times 3$ is equal to the product of 28 and 36.

$$(7 \times 4) \times (12 \times 3) = 28 \times 36$$

13) Choice D is correct.

$$20 = x \times 4 \rightarrow x = 20 \div 4 = 5$$

x equals to 5. Let's review the options provided:

A) $x + 4 \rightarrow 5 + 4 = 9$ 20 is not divisible by 9.

B) $2x - 4 \rightarrow 2 \times 5 - 4 = 6$ 20 is not divisible by 6.

C) $x - 2 \rightarrow 5 - 2 = 3$ 20 is not divisible by 3.

D) $x \times 4 \rightarrow 5 \times 4 = 20$ 20 is divisible by 20.

The answer is D.

14) Choice C is correct.

$$x - 12 = 18 \rightarrow x = 6$$
$$16 + y = 21 \rightarrow y = 5$$
$$x + y = 6 + 5 = 11$$

15) Choice D is correct.

$$\frac{5}{4} \times \frac{6}{2} = \frac{30}{8} = \frac{15}{4}$$

Choice D is equal to $\frac{15}{4}$.

$$\frac{5 \times 3}{4} = \frac{15}{4}$$

16) Choice B is correct.

$$5 + 3N = 41 \rightarrow 3N = 41 - 5 = 36 \rightarrow N = 12$$

17) Choice D is correct.

$$15 - 20 = -5$$

The temperature at midnight was 5 degrees below zero.

18) Choice B is correct.

Area of a triangle $= \frac{1}{2} \times (base) \times (height) = \frac{1}{2} \times 5 \times 8 = 20$

19) Choice D is correct.

$$area\ of\ a\ square = side \times side$$

Side

20) Choice A is correct.

First, find the pattern,

$$2 + 3 = 5 \rightarrow 5 + 4 = 9 \rightarrow 9 + 5 = 14 \rightarrow 14 + 6 = 20$$

The difference of two consecutive numbers increase by 1. The difference of 14 and 20 is 6.

So, the next number should be 27.

$$20 + 7 = 27$$

21) Choice B is correct.

$$average = \frac{sum\ of\ all\ numbers}{number\ of\ numbers} = \frac{6 + 10 + 12 + 23 + 45}{5} = 19.2$$

22) Choice B is correct.

There are 8 red ball and 20 are total number of balls. Therefore, probability that John will pick out a red ball from the basket is 8 out of 20 or $\frac{8}{8+12} = \frac{8}{20} = \frac{2}{5}$.

23) Choice C is correct.

An equilateral triangle has 3 lines of symmetry.

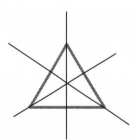

24) Choice B is correct.

$$10\ percent\ of\ 200 = 10\%\ of\ 200 = \frac{10}{100} \times 200 = 20$$

25) Choice C is correct.

Let's review the options provided:

A) $2 \times 2 = 4$ This is true!

B) $(4 + 1) \times 5 = 25$ This is true!

C) $6 \div (3 - 1) = 1 \rightarrow 6 \div 2 = 3$ This is NOT true!

D) $6 \times (4 - 2) = 12 \rightarrow 6 \times 2 = 12$ This is true!

26) Choice D is correct.

The shape has 6 equal sides. And is side is 4. Then, the perimeter of the shape is:

$$4 \times 6 = 24$$

27) Choice C is correct.

$$\frac{4}{5} - \frac{3}{5} = \frac{1}{5} = 0.2$$

28) Choice D is correct.

$N = 2$, then: $\frac{64}{2} + 4 = 32 + 4 = 36$

29) Choice D is correct.

Four people can paint 4 houses in 10 days. It means that for painting 8 houses in 10 days we need 8 people. To paint 8 houses in 5 days, 16 people are needed.

30) Choice B is correct.

Write the numbers in order: 4, 5, 8, 9, 13, 15, 18

Median is the number in the middle. Therefore, the median is 9.

31) Choice C is correct.

The biggest city is city C and the smallest city is city B.

Number of men in city A is 700 and number of men in city C is 300. Then: $700 - 300 = 400$

32) Choice B is correct

5,231.48245 rounded to the nearest tenth is 5,231.5

33) Choice A is correct

$$18a + 22 = 40$$

$$18a = 40 - 22$$

$$18a = 18$$

$$a = 1$$

34) Choice D is correct

All angles in a triangle sum up to 190 degrees. Two angles of a triangle measure 35 and 65.

$$35 + 65 = 100$$

Then, the third angle is: $180 - 100 = 80$

35) Choice C is correct

$$135 + 15 + 8 = 158$$

36) Choice D is correct

There are equal numbers of four types of cards. Therefore, the total number of cards must be divisible by 4. Only choice D (56) is divisible by 4.

37) Choice C is correct

15 seconds is one fourth of a minute. One fourth of 88 is 22.

$$88 \div 4 = 22$$

Jim types 22 words in 15 seconds.

38) Choice D is correct

There are equal

From the choice provided, only choice D is not equal to $\frac{2}{7}$. $\frac{12}{48} = \frac{1}{4}$

ISEE Lower Level Practice Test 1

Mathematics Achievement

1) Choice D is correct

Find common denominator and solve.

$$\frac{1}{5} + \frac{3}{4} = \frac{4}{20} + \frac{15}{20} = \frac{19}{20}$$

2) Choice D is correct

least common multiple (LCM) of 8 and 14 is the smallest number that is divisible by both 8 and 14. LCM = 56

3) Choice D is correct

The factors of 50 are: { 1, 2, 5, 10, 25, 50}

15 is not a factor of 50.

4) Choice D is correct

9 hour = 540 minutes. Write a proportion and solve.

$$\frac{90}{1} = \frac{540}{x} \quad \rightarrow \quad x = \frac{540}{90} = 6$$

5) Choice A is correct

8,923.2769 rounded to the nearest tenth is 8,923.3

6) Choice C is correct

One method to compare fractions is to convert them to decimals.

A. $\frac{3}{4}$ = 0.75

B. $\dfrac{2}{5}$ = 0.4

C. $\dfrac{7}{9}$ = 0.77

D. $\dfrac{2}{3}$ = 0.66

0.77 or $\dfrac{7}{9}$ is the largest number.

7) Choice D is correct

If 17 balls are removed from the bag at random, there will be one ball in the bag.

The probability of choosing a brown ball is 1 out of 18. Therefore, the probability of not choosing a brown ball is 17 out of 18 and the probability of having not a brown ball after removing 17 balls is the same.

8) Choice C is correct

The question is this: 1.75 is what percent of 1.25?

Use percent formula: part = $\dfrac{\text{percent}}{100}$ × whole

part= $\dfrac{\text{percent}}{100}$ ×1.25 ⇒ 1.75 = $\dfrac{\text{percent} \times 1.25}{100}$ ⇒175 = percent × 1.25 ⇒ percent = $\dfrac{175}{1.25}$ = 140

9) Choice B is correct

The question is this: 530.40 is what percent of 624?

Use percent formula: part = $\dfrac{\text{percent}}{100}$ × whole

530.40 = $\dfrac{\text{percent}}{100}$ × 624 ⇒ 530.40 = $\dfrac{\text{percent} \times 624}{100}$ ⇒53040 = percent × 624 ⇒

percent = $\dfrac{53040}{624}$ = 85

530.40 is 85 % of 624. Therefore, the discount is: 100% − 85% = 15%

10) Choice C is correct

Area of a triangle = $\frac{1}{2}$ (base)(height)

$A = \frac{1}{2}$ (28)(9) = 126

11) Choice A is correct

All angles in a triangle add up to 180 degrees.

$51 + 47 = 98$

$180 - 98 = 82$

12) Choice D is correct

$Perimeter = 2(width + length)$

$A = width \times length$

First, find the length of the rectangle.

$Perimeter = 2(width + length) \;\to\; 112 = 2(22 + length) \;\to\; 112$
$= 44 + 2(length) \to 68 = 2(length) \;\to\; length = 34$

$A = 22 \times 34 = 748$

13) Choice D is correct.

Since Julia running at 5.5 miles per hour and Julia is running at the speed of 6 miles per hour, each hour their distance decreases by 0.5 mile. So, it takes 15 hours to cover distance of 7.5 miles. $7.5 \div 0.5 = 15$

14) Choice B is correct

Since Julie gives 8 pieces of candy to each of her friends, then, then number of pieces of candies must be divisible by 8.

A. $187 \div 8 = 23.375$

B. $216 \div 8 = 27$

C. $343 \div 8 = 42.875$

D. $223 \div 8 = 27.875$

Only choice B gives a whole number.

15) Choice C is correct

$\$9 \times 10 = \$90$

Petrol use: $10 \times 2 = 20$ liters

Petrol cost: $20 \times \$1 = \$20$

Money earned: $\$90 - \$20 = \$70$

16) Choice C is correct

$0.025 \times 100 = 2.5\%$

17) Choice B is correct

$\dfrac{4}{100} = 0.04$

18) Choice D is correct

If the length of the box is 27, then the width of the box is one third of it, 9, and the height of the box is 3 (one third of the width). The volume of the box is:

$V = (length)(width)(height) = (27)(9)(3) = 729 \text{ m}^3$

19) Choice D is correct.

The failing rate is 25 out of 125 = $\dfrac{25}{125}$

Change the fraction to percent:

$$\frac{25}{125} \times 100\% = 20\%$$

20 percent of students failed. Therefore, 80 percent of students passed the exam.

20) Choice C is correct

$$\begin{array}{r} 37 \text{ hr.} \quad 25 \text{ min.} \\ - \ 23 \text{ hr.} \quad 38 \text{ min.} \\ \hline 13 \text{hr.} \quad 47 \text{min.} \end{array}$$

21) Choice A is correct

An obtuse angle is an angle of greater than 90 degrees and less than 180 degrees. Only choice a is an obtuse angle.

22) Choice A is correct

Let x be the number of raptors Daniel saw on Monday. Then:

$$Mean = \frac{x+9+14+12+5}{5} = 10 \rightarrow x + 40 = 50 \qquad \rightarrow \qquad x = 50 - 40 = 10$$

23) Choice D is correct.

There are 10 squares and 6 of them are shaded. Therefore, 6 out of 10 or $\frac{6}{10} = \frac{3}{5}$ are shaded.

24) Choice B is correct

Write a proportion and solve. $\frac{2}{3} = \frac{x}{90}$

Use cross multiplication: $3x = 180 \rightarrow x = 60$

25) Choice D is correct

To find the discount, multiply the number by (100% − rate of discount).

Therefore, for the first discount we get: $(500)(100\% - 25\%) = (500)(0.75)$

For the next 15 % discount: $(500)(0.75)(0.85)$

26) Choice D is correct.

A quadrilateral with one pair of parallel sides is a trapezoid.

27) Choice A is correct.

Plug in 20 for A in the equation. Only choice A works. $A + 20 = 40$

$$20 + 20 = 40$$

28) Choice D is correct

$1 \; hour$: \$4.75

$8 \; hours$: $8 \times \$4.75 = \$38$

29) Choice C is correct.

There are 44 students in the class. 18 of the are male and 26 of them are female.

26 out of 44 are female. Then:

$$\frac{26}{44} = \frac{x}{100} \rightarrow 2{,}600 = 44x \rightarrow x = 2{,}600 \div 44 \approx 59\%$$

30) Choice B is correct.

1 pizza has 8 slices. 18 pizzas contain (18×8) 144 slices.

ISEE Lower Level Practice Test 2

Quantitative Reasoning

1) Choice D is correct.

$$\frac{8}{2} - \frac{3}{2} = \frac{5}{2} = 2.5$$

2) Choice B is correct.

$$48 = 3 \times N + 12 \rightarrow 3N = 48 - 12 = 36 \rightarrow N = 12$$

3) Choice C is correct.

Four times a number N is $4 \times N$. When 3 is added to it, the result is:

$$3 + (4 \times N) = 23 \rightarrow 4N + 3 = 23$$

4) Choice D is correct.

78 divided by 5, the remainder is 3. 45 divided by 7, the remainder is also 3.

5) Choice A is correct.

$$2,400 - 606 = 1,794$$

6) Choice C is correct.

All angles in a triangle sum up to 180 degrees. The triangle provided is an isosceles triangle. In an isosceles triangle, the three angles are 45, 45, and 90 degrees. Therefore, the value of x is 45.

7) Choice B is correct.

$$5 \text{ percent of } 480 = \frac{5}{100} \times 480 = \frac{1}{20} \times 480 = \frac{480}{20} = 24$$

8) Choice B is correct.

The ratio of red marbles to blue marbles is 3 to 2. Therefore, the total number of marbles must be divisible by 5.

$3 + 2 = 5$

32 is the only one that is not divisible by 5.

9) Choice D is correct.

$$\textit{Area of a square } = \textit{ side} \times \textit{side} = 81 \rightarrow \textit{side} = 9$$
$$\textit{Perimeter of a square } = 4 \times \textit{ side} = 4 \times 9 = 36$$

10) Choice B is correct.

$$5 + 3 = 8, \quad 8 + 4 = 12, \quad 12 + 5 = 17, \quad 17 + 6 = 23$$

11) Choice D is correct.

The length of the rectangle is 18. Then, its width is 6.

$$18 \div 3 = 6$$

$$\textit{Perimeter of a rectangle} = 2 \times \textit{width} + 2 \times \textit{length} = 2 \times 6 + 2 \times 18 = 12 + 36$$
$$= 48$$

12) Choice C is correct.

$$\textit{Mary's Money} = y$$

$$\textit{John's Money} = y + 10$$

$$\textit{John gives Mary } \$12 \rightarrow y + 10 - 12 = y - 2$$

13) Choice D is correct.

Dividing 107 by 6 leaves a remainder of 5.

14) Choice C is correct.

$$6{,}000 + A - 200 = 7{,}400 \rightarrow 6{,}000 + A = 7{,}400 + 200 = 7{,}600 \rightarrow A = 7{,}600 - 6{,}000$$
$$= 1{,}600$$

15) Choice D is correct.

$75 off is the same as 15 percent off. Thus, 15 percent of a number is 75.

Then: $15\% \textit{ of } x = 75 \rightarrow 0.15x = 75 \rightarrow x = \dfrac{75}{0.15} = 500$

16) Choice A is correct.

$\frac{3}{2} = 1.5$, the only choice provided that is less than 1.4 is choice A.

$$\frac{3}{2} = 1.5 > 1.4$$

17) Choice A is correct.

$$x + 3 = 6 \rightarrow x = 3$$
$$2y = 8 \rightarrow y = 4$$
$$y - x = 4 - 3 = 1$$

18) Choice C is correct.

$$310 - x + 116 = 225 \rightarrow 310 - x = 225 - 116 = 109 \rightarrow x = 310 - 109 = 201$$

19) Choice C is correct.

25 percent of $44.00 is $11. (Remember that 25 percent is equal to one fourth)

20) Choice C is correct.

$$8.08 - 5.6 = 2.48$$

21) Choice D is correct.

$$500 + \square - 180 = 1,100 \rightarrow 500 + \square = 1,100 + 180 = 1,280$$

$$\square = 1,280 - 500 = 780$$

22) Choice D is correct.

$\frac{1}{6}$ of students are girls. Therefore, $\frac{5}{6}$ of students in the class are boys. $\frac{5}{6}$ of 60 is 50. There are 50 boys in the class. $\frac{5}{6} \times 60 = \frac{300}{6} = 50$

23) Choice A is correct.

$$N \times (5 - 3) = 12 \rightarrow N \times 2 = 12 \rightarrow N = 6$$

24) Choice C is correct.

If $x \blacksquare y = 3x + y - 2$, Then:

$$4 \blacksquare 12 = 3(4) + 12 - 2 = 12 + 12 - 2 = 22$$

25) Choice C is correct.

Of the numbers provided, 0.8923 is the greatest.

26) Choice A is correct.

$$\frac{7}{8} - \frac{3}{4} = \frac{7}{8} - \frac{6}{8} = \frac{1}{8} = 0.125$$

27) Choice A is correct.

The closest number to 4.02 is 4.

28) Choice D is correct.

$14 \div (2 + 5) = 14 \div 7 = 2$ not 5

29) Choice B is correct

240 + 35 = 275

30) Choice B is correct

Let's order number of shirts sold per month:

$$130, 140, 145, 150, 160, 170$$

Median is the number in the middle. Since, there are 6 numbers (an even number) the

Median is the average of numbers 3 and 4: Median is: $\frac{145+150}{2} = 147.5$

31) Choice D is correct

$180 \div 20 = 9$

32) Choice B is correct

Speed $= \dfrac{distance}{time}$

$68 = \dfrac{2,600}{time}$ → $time = \dfrac{2,600}{68} = 38.23$

It takes Nicole about 38 hours to go from city A to city B.

33) Choice C is correct

$\dfrac{13}{25} = 0.52$

34) Choice B is correct

$12a + 20 = 140$

$12a = 140 - 20$

$12a = 120$

$a = 10$

35) Choice B is correct

Thousandths

36) Choice D is correct

6 is NOT a prime factor. (it is divisible by 2 and 3)

37) Choice C is correct

$$\frac{25}{1,000} = 0.025$$

38) Choice D is correct

$\frac{1}{3}$ of 24 hours is 8 hours. $\frac{1}{3} \times 24 = \frac{24}{8} = 8$

ISEE Lower Level Practice Test 2

Mathematics Achievement

1) Choice A is correct

Sixty-five thousand, eight hundred nineteen is written as 65,819.

2) Choice C is correct

The factors of 18 are: { 1, 2, 3, 6, 9, 18}

The factors of 32 are: { 1, 2, 4, 8, 16, 32}

greatest common factor (GCF) = 2

3) Choice A is correct

From choices provided, only 12 is NOT a multiple of 5.

4) Choice C is correct

Use the Pythagorean Theorem to find the length of the third side:

$a^2 + b^2 = c^2$

$6^2 + 8^2 = c^2$

$100 = c^2$

$c = 10$

The perimeter of the triangle is: 6 + 8 + 10 = 24

5) Choice D is correct

The name of a rectangle with sides of equal length is square.

6) Choice C is correct.

The question is that number 56,743.21 is how many times of number 5.674321. The answer is 10,000.

7) Choice B is correct.

$\frac{3}{11}$ means 3 is divided by 11. The fraction line simply means division or ÷.

Therefore, we can write $\frac{3}{11}$ as $3 \div 11$.

8) Choice C is correct.

Lily eats 2 pancakes in 1 minute $\Rightarrow$ Lily eats 2×5 pancakes in 5 minutes.

Ella eats 2 ½ pancakes in 1 minute $\Rightarrow$ Ella eats 2 ½ × 5 pancakes in 5 minutes.

In total Lily and Ella eat 10 + 12.5 = 22.5 pancakes in 5 minutes.

9) Choice A is correct.

Simplify each choice provided using order of operations rules.

A. $8 - (-2) + (-18) = 8 + 2 - 18 = -8$
B. $2 + (-3) \times (-2) = 2 + 6 = 8$
C. $-6 \times (-6) + (-2) \times (-12) = 36 + 24 = 60$
D. $(-2) \times (-7) + 4 = 14 + 4 = 18$

Only choice A is -8.

10) Choice C is correct

$0.87 + 1.4 + 3.23 = 5.5$

11) Choice C is correct

Perimeter of a rectangle = 2(*length* + *width*) = 2(8 + 5) = 26

12) Choice D is correct

To solve this problem, divide $6\frac{1}{2}$ by $\frac{1}{4}$.

$$6\frac{1}{2} \div \frac{1}{4} = \frac{13}{2} \div \frac{1}{4} = \frac{13}{2} \times \frac{4}{1} = 26$$

13) Choice C is correct.

Subtract hours: 2 − 6 = − 4

Subtract the minutes: 15 − 30 = − 15

The minutes are less than 0, so:

Add 60 to minutes (−15 + 60 = 45 minutes)

Subtract 1 from hours (−4 − 1 = −5) the hours are less than 0, add 24: (24 − 5 = 19)

The answer is 19:45 that is equal to 7:45

14) Choice B is correct.

$Area \;=\; width \times height$

$Area \;=\; 72$

$Width = 8,\; 72 = 8 \times height$

$height = \dfrac{72}{8} = 9$

15) Choice C is correct.

$27{,}980 \;\leq\; 28{,}097 \;\leq\; 28{,}207 \;\leq\; 29{,}700$

16) Choice B is correct.

Low temperature is 24°f cooler than the temperature at 12:00 PM that is 76°f, that means low temperature is 52°f (76°f – 24°f).

17) Choice C is correct

$(5 + 7) \div (3^2 \div 3) = (12) \div (6 \div 3) = (12) \div (2) = 6$

18) Choice D is correct

$2.26 + 14.69 + 2.50 + 4.66 + 17.99 = 42.1$

19) Choice D is correct

The line is from 1 to -8. $1 - (-8) = 1 + 8 = 9$

20) Choice A is correct.

To find total number of students in Riddle Elementary school, add number of all students.

$86 + 32 = 118$

21) Choice A is correct.

$\frac{11}{19} \cong 0.579$

22) Choice D is correct.

$80 \times 34 = 2,720$

23) Choice C is correct.

The first picture is divided to 16 parts that 6 parts of it is shaded ($\frac{6}{16}$). The second picture is divided to 10 parts that 5 parts of that is shaded ($\frac{5}{10}$).

24) Choice C is correct.

The digit 6 has a value of $6 \times \dfrac{1}{100}$

The digit 4 has a value of $4 \times 10,000$

The digit 8 has a value of $8 \times 10 = 800$

The digit 5 has a value of $5 \times 1,000$

25) Choice B is correct.

Three – digit odd numbers that have a 6 in the hundreds place and a 3 in the tens place are 631, 633, 635, 637, 639. 637 is one of the choices.

26) Choice B is correct

$$average \ (mean) = \frac{sum \ of \ terms}{number \ of \ terms} = \frac{9+12+15+16+19+16+14.5}{7} = 14.5$$

27) Choice A is correct

$$probability = \frac{desired \ outcomes}{possible \ outcomes} = \frac{4}{4+3+7+10} = \frac{4}{24} = \frac{1}{6}$$

28) Choice C is correct.

0.62 is equal to $\dfrac{62}{100}$

29) Choice B is correct.

Find the least common denominator (LCD), then rewriting each term as an equivalent fraction with the LCD. Then we compare the numerators of each fraction and put them in correct order from least to greatest or greatest to least.

LCD of 2, 8, 4 and 16 is 16. Rewrite the input fractions as equivalent fractions using the LCD:

A. $\dfrac{8}{16}$ B. $\dfrac{6}{16}$ C. $\dfrac{12}{16}$ B. $\dfrac{9}{16}$

So, choice B has the least value.

30) Choice D is correct.

$1 \ year = 365 \ days, 1 \ day = 24 \ hours$

$1 \ year = 365 \times 24$

$1 \ year = 8,760$

"Effortless Math" Publications

Effortless Math authors' team strives to prepare and publish the best quality Mathematics learning resources to make learning Math easier for all. We hope that our publications help you or your student Math in an effective way.

We all in Effortless Math wish you good luck and successful studies!

Effortless Math Authors

www.EffortlessMath.com

... So Much More Online!

✓ FREE Math lessons

✓ More Math learning books!

✓ Mathematics Worksheets

✓ Online Math Tutors

Need a PDF version of this book?

Visit www.EffortlessMath.com

41346251R00141

Made in the USA
San Bernardino, CA
02 July 2019